Two Manuals Combined into One Book

AF812689

Religious Support

And

Marine Combat Stress

Department of Defense

Pentagon
Publishing

Field Manual
No. 1-05

Headquarters
Department of the Army
Washington, DC, 18 April 2003

Religious Support

Contents

Distribution Restriction. Approved for public release; distribution is unlimited.

*This publication supersedes FM 16-1, 26 May 1995.

Preface

The Chaplain Corps is the Army's primary agency for practically ensuring the free exercise of religion for America's soldiers. This responsibility also applies to the Chaplain Corps' religious support to Army family members, Department of Army (DA) civilians, and other authorized personnel. The Chaplain Corps ensures free exercise of religion by executing the command religious programs of Army commanders.

PURPOSE

FM 1-05 establishes the Army's definitive doctrine for the execution of comprehensive religious support across the full spectrum of operations. It is founded upon the Army's operational doctrine articulated in FM 3-0 and FM 101-5 and other supporting documents. FM 1-05 is compatible with the joint religious support doctrine contained in JP 1-05. This doctrine pronounces that the personal delivery of religious support is the primary focus of the Chaplain Corps and places greater emphasis on the chaplain's role as an advisor to the commander. It takes into account emerging technology, possesses a broader focus on religious support at echelons above brigade, and speaks clearly to the interface between religious support operations and civil military operations. It is the authoritative, but not prescriptive, guide to the missions, roles, responsibilities, and duties the Chaplain Corps must execute to deliver comprehensive religious support.

APPLICABILITY

FM 1-05 provides guidance to commanders at all echelons responsible for providing command religious support programs to their units and organizations, as well as, to Chaplains, Chaplain Assistants and other personnel responsible to plan, prepare, and provide religious support to The Army. This guidance is the principle tool for professional education in the art of religious support operations. Trainers at all echelons should employ it as the authoritative guide for the planning and execution of religious support training. For overseas theaters, this doctrine applies to U.S. unilateral operations only, subject to applicable host nation laws and agreements.

USER INFORMATION

The proponent for this publication is the U.S. Army Training and Doctrine Command (TRADOC). Send comments and recommendations on DA Form 2028 (Recommended Change to Publications and Blank Forms) to Commandant, U.S. Army Chaplain Center and School, ATTN: ATSC-DCD, Fort Jackson, SC 29207.

Unless this publication states otherwise, masculine nouns and pronouns do not refer exclusively to men.

Chapter 1

RELIGIOUS SUPPORT FUNDAMENTALS

CONTENTS

The First Amendment of the United States Constitution reinforced by Title 10 U.S. Code (USC) and Army Regulation (AR) 165-1, guarantees every American the right to the free exercise of religion. Commanders are responsible for those religious freedoms within their command. On behalf of the commander, Chaplains provide and perform Religious Support (RS) in the Army to ensure the free exercise of religion. Chaplains are obligated to provide for those religious services or practices that they cannot personally perform. Chaplains perform religious support when their actions are in accordance with the tenets or beliefs of their faith group. Chaplain assistants assist the chaplain in providing or performing this religious support. Religious support includes—

- Religious services.
- Rites.
- Sacraments.
- Ordinances.
- Pastoral care.
- Religious education.
- Family life ministry.
- Institutional ministry.
- Professional support to the command and staff.
- Management and administration.
- Religious/humanitarian support.
- Religious support planning/operations and religious support training.

1-1. The Chaplain's statutory (10 USC) and regulatory (AR 165-1) understanding of religious support includes—

- Provide and/or perform worship services, rites, sacraments, ordinances, pastoral and spiritual care and religious education to nurture the living, care for the dying, and honor the dead.

- Advice to the commander on morals and morale as affected by religion, the impact of local religions on the military mission, and the ethical impact of command decisions.

1-2. Examples of religious support include—

- Worship, in accordance with one's faith, mindful of Army regulations and mission.
- Religious guidance, care, and counsel.
- Observation of Holy days and observances.
- Participation in rites, sacraments, and/or ordinances of one's faith.
- Practice of dietary laws.
- Medical treatment according to one's religious belief.
- Resolution of medical treatment challenges, religious apparel issues and dietary restrictions arising from one's religious belief. (See AR 600-20 for further guidance on religious accommodation.)

1-3. The Unit Ministry Team (UMT) is a task-organized, mission-based team designed to support the religious, spiritual, and ethical needs of soldiers and their families, members of other services, and authorized civilians.

1-4. The UMT consists of at least one chaplain and one chaplain assistant. UMTs are identified according to either a Modified Table of Organization and Equipment (MTOE) or a Table of Distribution and Allowances (TDA). Certain roles, functions and expectations of these UMTs depend on whether they are authorized by the MTOE or by the TDA. Continued mission and task analysis determines the distribution and size of UMT assets. The UMT is embedded into the Army Force Structure as organic and central to the organization.

1-5. Religious Support in the Army has no civilian equivalent. Chaplains execute their distinct religious support mission for soldiers, members of other military services, family members and/or authorized civilians in a variety of geographical locations, operational situations, and circumstances.

1-6. The religious support model in Figure 1-1 depicts supporting functions and tasks the UMT performs or provides for the free exercise of religion, as stated in 10 USC.

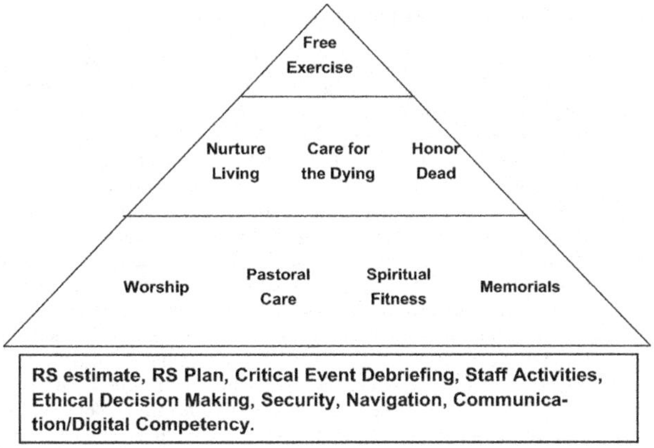

Figure 1-1. Religious Support Model

> Religious Support in the Army has no civilian equivalent.

HISTORICAL FOUNDATIONS

1-7. Chaplains have served in the U.S. Army since the Revolutionary War. The Continental Congress enacted regulations and salaries governing chaplains. George Washington, as Commander-in-chief, was resolute in his view that chaplains were necessary for the good ordering of the military and steadfast in his conviction that only well-qualified men should be chosen to serve as chaplains. To Colonel George Taylor he wrote that chaplains should be men of "Character and good conversation...who will influence the manners of the Corps both by precept and example." On 29 July 1775, the Continental Congress provided for the appointment of chaplains for the armed forces. That date is the recognized "birthday" of the Chaplain Corps, making it the second oldest branch in the Army, preceded only by the Infantry. 10 USC , Sections 3073, 3547, and 3581 codify the position of Chaplain in the Army. The 10 USC and AR 165-1 prescribe the duties of chaplains and require commanders to provide for the religious needs of soldiers. Religious support to families and authorized civilians is a derived task of this duty.

1-8. The War Department established the position of chaplain assistant by General Order Number 253, on 28 December 1909. The commanding officer detailed one enlisted soldier to special duty to any organization in which a chaplain was assigned. This detailed soldier assisted the chaplain in the performance of his official duties. The Army established the military occupational specialty (MOS) chaplain assistant as an official MOS in 1968.

THE CHAPLAIN'S AUTHORITY FOR MINISTRY

1-9. The government recognizes that the primary role of religious support belongs to the religious community. The federal government also recognizes the special religious support needs of armed forces personnel and the difficulty of providing organized, effective religious support. The isolation, unique environment, and training for combat zones throughout the world generate special requirements for religious support. The federal government provides the chaplaincy to meet these needs. Chaplains serve in the Army as clergy representing the respective faiths or denominations that endorse them. A chaplain's call, ministry, message, ecclesiastical authority, and responsibility come from the religious organization that the chaplain represents. Chaplains preach, teach, and conduct religious services, in accordance with the tenets and rules of their tradition, the principles of their faith, and the dictates of conscience. They also perform ecclesiastical rites and ceremonies and administer the sacraments and ordinances of their faith community. When a conflict arises between the standards of a chaplain's faith tradition and the requirements of a religious support mission, chaplains are required to provide for the religious needs of the soldier by obtaining other chaplains or qualified people to perform the needed religious support.

1-10. Chaplains, ordained and endorsed by distinctive faith groups, must be people of integrity who consistently live according to the highest principles of the Army and the distinctive faith groups they represent. Chaplain assistants are also expected to be people of integrity whose lives reflect high moral standards and the values of the Army. The UMT must mirror the Army's values – Loyalty, Duty, Respect, Selfless Service, Honor, Integrity, and Personal Courage.

1-11. As religious leaders, chaplains continue to meet the vision of General George Washington by setting the standard for moral and ethical behavior by upholding the Army Chaplaincy Values: Spirituality, Accountability, Compassion, Religious Leadership, Excellence, and Diversity (SACRED).

Army Chaplaincy Values
Spirituality
The bedrock of all faiths, providing meaning and direction in life.
Accountability
The means to measure how we practice what we preach.
Compassion
The love that sees beyond the surface and discerns our common humanity.
Religious leadership
Interprets the role of leadership prophetically.
Excellence
The standard for staff work and religious responsibilities.
Diversity
Being respectful of different views and ideas that are not like our own.

RELIGIOUS SUPPORT MISSION

1-12. The mission of the UMT is to provide and perform religious support to soldiers, families, and authorized civilians as directed by the commander. Chaplains serve as personal staff officers to commanders at all levels of the command providing essential information on troop and unit morale, quality of life matters, free exercise of religion issues, ethical decision-making, and the impact of religion on the operation.

RELIGIOUS SUPPORT FUNCTIONS

1-13. Comprehensive religious support includes the following three major functions:

- ***Nurture the Living.*** In preparation for missions that span the full spectrum of operations UMTs develop and provide religious support activities to strengthen and sustain the spiritual resilience of soldiers and family members. During the battle, UMTs bring hope and strength to those who have been wounded and traumatized in body, mind, and spirit, assisting the healing process.
- ***Care for the Dying.*** UMTs provide religious support, spiritual care, comfort, and hope to the dying. This focus of religious support affirms the sanctity of life, which is at the heart of the chaplaincy. Through prayer and presence, the UMT provides the soldier with courage and comfort in the face of death.
- ***Honor the Dead.*** Our Nation reveres those who have died in military service. Religious support honors the dead. Funerals, memorial services, and ceremonies, reflect the emphasis the American people place on the worth and value of the individual. Chaplains conduct these services and ceremonies, fulfilling a vital role in rendering tribute to America's sons and daughters who paid the ultimate price serving the Nation in the defense of freedom.

RELIGIOUS SUPPORT ACTIVITIES

1-14. Commanders provide opportunities for the free exercise of religion. UMTs provide and/or perform the following major religious activities, within the unique military context, in support of the command religious plan:

- ***Religious Services:*** Worship; Funerals; Memorial Services and Ceremonies; Prayer Breakfasts.
- ***Rites, Sacraments, Ordinances:*** Marriages; Burials; Baptisms; Confirmations; Blessings; Daily Prayers; other ministrations and religious requirements.
- ***Pastoral Care/Counseling:*** Taking part in command activities; visiting soldiers; calling on families; activities and unit "ministry of presence"; individual and group pastoral counseling; and similar pastoral activities. This may also include elements of the spiritual fitness training program. .
- ***Religious Education:*** Study of religious teachings including activities of faith sustainment (enrichment, formation, study or instruction), marriage enrichment, problem solving, communication skills, parenting skills, youth programs, religious literature.

- ***Family Life Ministry:*** Training UMTs, directing the Chaplain Family Life Center (CFLC). Family Life Ministry is implemented through the CFLC programs and formal training.
- ***Institutional Ministry:*** Preserving soldiers' fighting strength by maintaining their health and supporting family members receiving medical services. Institutional ministry occurs in hospitals and confinement or correction facilities.
- ***Professional Support to the Commander and Staff:*** Support to the commander on matters of religion, morals, and morale, as affected by religion and the impact of indigenous religions on the military mission. Assesses the unit climate by visiting soldiers and providing advice on the religious needs of soldiers and families.
- ***Management and Administration:*** Beyond the normal administration requirements and actions required by the Command Religious Plan, the UMT manages funds, facilities, manpower, supplies, and equipment.
- ***Religious/Humanitarian Support:*** Support programs to Joint Military Operations. Such support includes humanitarian support programs on issues of religion, morale, morals, and ethics.
- ***Religious Support Planning/Operations:*** Providing commanders with Religious Support plans and programs for all mobilization requirements spanning the full spectrum of military operations, as well as, planned ministries for families of deployed soldiers.
- ***Religious Support Training:*** Training that includes spiritual fitness (Spiritual fitness training is the training and development of personal qualities needed to sustain a soldier in times of danger, hardship, and tragedy. These qualities are based in religious, philosophical and human values and they form the basis of character, decision-making and integrity.), moral leadership (see AR 165-1, Chapter 11), soldier and family member suicide prevention, violence in the work place, professional leadership, and training required by UMTs, commanders, staff, soldiers and family members.

ROLES AND RESPONSIBILITIES

1-15. Commanders, chaplains, and chaplain assistants have distinct roles and responsibilities in providing and performing religious support to soldiers, their family members and authorized civilians.

COMMANDER

1-16. The commander provides for the free exercise of religion of soldiers and families thereby enhancing the spiritual health of the command. To assure readiness across the full spectrum of operations, commanders provide for a wide range of religious support activities. Commanders use the Commanders' Religious Program (CRP) to resource the religious support plan.

1-17. The commander accommodates religious practices, in accordance with Army regulations (See AR 600-20). The Army policy is to accommodate religious practices when those accommodations will not have an adverse impact on military readiness, unit cohesion, standards, health, safety or discipline, or otherwise interfere with the performance of the soldier's military duties. Accommodation of a soldier's religious practices must be weighed against military necessity and cannot be guaranteed at all times.

CHAPLAIN

1-18. Chaplains personally deliver religious support. They have two roles: religious leader and staff officer.

- The chaplain as a religious leader executes the religious support mission, which ensures the free exercise of religion for soldiers and authorized personnel.
- The chaplain is a non-combatant and will not bear arms (AR 165-1).

> The Chaplain is a non-combatant and will not bear arms in accordance with AR 165-1.

1-19. The chaplain is a personal staff officer responsible for coordinating the religious assets and operations within the command. The chaplain is a confidential advisor to the commander for religious matters. A chaplain is located at every echelon of command above battalion. Besides his common staff responsibilities, the chaplain's specific responsibilities include —

- Advise the commander on issues of religion, ethics, and morale (as affected by religion), including the religious needs of all personnel for whom the commander is responsible.
- Provide commanders with pastoral care, personal counseling, advice, confidentiality, and sacred confidence.
- Develop and implement the commander's religious support program.
- Exercise staff supervision and technical control over religious support throughout the command.
- Provide moral and spiritual leadership to the command and community.
- Coordinate religious support with unit ministry teams of higher and adjacent headquarters, other services, and multinational forces or coalition partners.
- Translate operational plans into battlespace ministry priorities for religious support.
- Help the commander ensure that all soldiers have the opportunity to exercise their religion.
- Advise the commander and staff, with the G5 (S5), on the impact of the faith and practices of indigenous religious groups in an area of operation.
- Perform or provide religious rites, sacraments, ordinances, services, and pastoral care and counseling to nurture the living, care for the dying and honor the dead.
- Provide religious support to the command and community to include confined or hospitalized personnel, enemy prisoners of war (EPWs), civilian detainees, and refugees.
- Support the commander in the execution of Civil Military Operations (See appendix A).

1-20. The Chaplain trains, equips and supports the subordinate chaplain and chaplain assistant (FM 101-5).

1-21. As a minimum, the chaplain requires a SECRET clearance. This will allow the chaplain to have access to the command post and enable him/her to be part of the planning process.

1-22. Although chaplains do not have command or Uniform Code of Military Justice (UCMJ) authority, they do exercise staff supervision and functional direction of religious support personnel and activities.

1-23. The proper title for a chaplain is "chaplain" regardless of military rank and professional title.

CHAPLAIN ASSISTANT

1-24. The chaplain assistant (CA) is a combatant, trained in religious support and staff skills; who possesses a SECRET clearance; manages force protection, provides security for the UMT on the battlefield, and accompanies the chaplain in the area of operation. The CA is a soldier trained to assist the chaplain in providing religious support. Under the direction of the chaplain, the CA coordinates and synchronizes religious support operations. In the absence of the chaplain, the CA continues the religious support mission for the commander as a staff representative, but does not assume the religious leadership role of the Chaplain.

1-25. Unit Ministry Team Chaplain Assistant. Under the supervision of the chaplain, the CA—

- Assists in religious support planning and operations.
- Coordinates the UMT movement plan, monitors the tactical situation, and attends briefings.
- Coordinates and prepares for religious activities.
- Coordinates for and conducts equipment maintenance.
- Assesses soldier morale and provides this information to the chaplain.
- Conducts pre-counseling interviews.
- Assists with combat operations stress response (COSR), identification, intervention, and care.
- Performs tactical, logistical, and administrative support for the UMT.
- Requires access to classified messages, systems, and secure areas to accomplish the religious support mission.
- When an NCO, the CA serves as a liaison for the chaplain to the noncommissioned officers (NCOs) of the command.

> Chaplain Assistants require a SECRET clearance to accomplish the religious support mission.

1-26. The Chaplain Assistant NCO supervises, mentors, and trains chaplain assistants and serves as the religious support section noncommissioned officer in charge (NCOIC) according to the echelon assigned. In addition to executing the roles and functions of a CA of lesser rank, the NCOIC performs many of the following staff functions:

- Participates in battle staff planning, tracking, and execution.
- Procures contracts for goods and services, and coordinates and monitors the CRP.

- Provides training in suicide prevention awareness, intervention for COSR, and performance counseling.
- Plans, briefs, and performs operational training with other staff NCOs.
- Coordinates taskings.
- Serves as advisor to the unit CSM and other key NCO leaders.
- Coordinates religious support during joint, multinational, and interagency operations.

ORGANIZATION

UNIT MINISTRY TEAM AND MODIFIED TABLE OF ORGANIZATION AND EQUIPMENT (MTOE) UNITS

1-27. UMTs authorized by the Modified Table of Organization and Equipment (MTOE) are assigned to units whose primary mission is warfighting. The UMT deploys with its unit and provides religious support for all units in the commander's area of responsibility during each stage of force projection.

1-28. At echelons above brigade, the senior chaplain is at times listed in the command section of the MTOE. Additional chaplains, chaplain assistants, and UMT equipment are listed in a separate section of the MTOE.

1-29. Comprehensive religious support requires the assignment and deployment of UMTs in maneuver, maneuver support, and maneuver sustainment units at all echelons, beginning at the battalion. UMTs move forward to provide religious support to all elements of the squadron/battalion: companies, troops, batteries, platoons, squads, and teams.

1-30. In operations where lethality and intensity are high, religious support becomes increasingly important to soldiers. UMTs provide hope, faith, encouragement, and compassion to soldiers experiencing shock, fatigue, isolation, fear, and death. In the chaos and uncertainty of military operations, the UMT is a reminder of the presence of God.

UNIT MINISTRY TEAMS AND TABLE OF DISTRIBUTION AND ALLOWANCES (TDA) UNITS

1-31. UMTs authorized by a TDA are assigned to units whose primary missions vary widely. The missions and responsibilities associated with TDA organizations depend upon the type of organization served. TDA organizations include installations, schools, hospitals, disciplinary barracks, some Major Army Commands (MACOMs), and Headquarters, Department of the Army (HQDA). The Installation UMT provides the training base for all MTOE and TDA UMTs. In addition to providing religious support directly to the units, the TDA UMTs manage, supervise, and support the religious support mission of their organization.

PROVIDING UNIT RELIGIOUS SUPPORT

1-32. Unit religious support is provided in the following three ways:

- ***Unit support:*** support provided to the unit to which the UMT is assigned or attached. The team normally gives first priority to this mission.

- ***Area support:*** support provided to soldiers, members of other services, and authorized civilians who are not a part of the team's unit, but operate within the supporting unit's area of operations (AO).
- ***Denominational support:*** support given to soldiers and other authorized persons of the chaplain's denomination or distinctive faith group. Limited assets affect the availability of denominational support.

1-33. The UMT extends opportunities for soldiers to serve as volunteers in worship, religious education, and service to the unit. The sponsoring UMT is responsible for recruiting, training, and supervising volunteers. Under the supervision of the unit chaplain, soldiers may serve as members of their own faith groups as Distinctive Faith Group Leaders (DFGL). Denominational certification, the approval of the commander, and the final approval of the MACOM/State Area Command (STARC)/United States Army Reserve Command (USARC) or Major Subordinate Command (MSC) chaplain are required before a soldier may serve as a DFGL. (See AR 165-1)

SUMMARY

1-34. The Chaplain Corps and Regiment remains committed to providing for the fundamental freedom of the free exercise of religion as guaranteed by the Constitution. This is accomplished by—

- Performing religious activities according to the chaplain's faith tradition.
- Providing professional and effective advice to the commander, in coordination with commander's staff, regarding the needs of soldiers not of the chaplain's faith tradition.
- Developing effective teamwork within the UMT.
- Through cooperation with other chaplains without compromising the tenets of one's own faith.

1-35. Through it all, the soldier's spiritual health and welfare, as well as the soldier's spiritual freedoms, remain the focal point. Thus the Chaplain Corps' unofficial motto: *"Bringing God to Soldiers, and Soldiers to God."*

RESOURCES

1-36. The following resources were used for writing and updating this doctrine.

- AR 165-1.
- AR 600-20.
- JP 1-05.
- JP 5-00.2.
- JP 3-0.
- FM 1 (100-1).
- FM 3-0 (100-5).
- FM 16-1.
- FM 22-100.
- DA Pam 165-3.
- Military Rules of Evidence 503 UCMJ.
- U.S. Army Chaplaincy Strategic Plan FY 2000-FY 2005.
- Chaplain Support Force XXI White Paper.
- Chaplain Assistant Concept Paper, 2000.

Chapter 2

RELIGIOUS SUPPORT ENVIRONMENT

"We have to remember something simple but important: Making history is messy. Studying Normandy today looks like arrows and unit symbols. Normandy on 6 June 1944 looked like Saving Private Ryan *– dangerous and chaotic."*

MG James Dubik

The mission and operational environment determine the shape of religious support. The team conducts religious support effectively with a clear understanding of the context and complexities of military operations in which religious support is provided and performed.

CONTENTS

RELIGIOUS SUPPORT AND THE THREAT ENVIRONMENT

2-1. With the collapse of the Soviet Union, the bilateral world of the post WWII era, pitting the political ideologies of communism and capitalism against one another, came to an end. As a consequence, the operational climate and the organizational structure of the U.S. Army began to change. The Army moved away from the concept of a forward deployed Army based overseas to the concept of a force projection Army based in the United States. "Because the Army is a power projection force, its units deploy from and return to home stations." (FM 3-0, pg. 4-9) The changes in operational climate, organization, and deployment posture require adaptations in the methods of providing religious support.

THE THREAT

"The enemies we are likely to face through the rest of the decade and beyond will not be 'soldiers,' with the disciplined modernity the term conveys, but 'warriors' – erratic primitives of shifting allegiance, habituated to violence, with no stake in civil order."

LTC Ralph Peters, Fighting for the Future

2-2. Potential threats to U.S. forces are many and varied. They range from a local populace with small arms to a conventional force possessing weapons of mass destruction (WMD). Global, regional, and local challenges confront the U.S. Army with adversaries who, as a result of their religious beliefs, may interpret such fundamental ideas as right and wrong, the value of human life, or even the concept of victory and defeat very differently.

2-3. Failing states and the associated fragmentation due to religious or ethnic strife will continue to promote regional instability and produce security concerns for the United States and our allies. The security environment is further complicated by challenges that transcend state borders, such as ethnic disputes and religious rivalries.

2-4. Threats can be of short duration or protracted over a long period of time. They can be waged over competing ideology, ethnic identity, economic policies, national aspirations, religious conflict or any number of social and political issues. Struggles can be waged by a variety of means: diplomatic, informational, military, or economic.

2-5. Asymmetric threats (unconventional or inexpensive approaches) to the United States (such as terrorism—both domestic and international, guerilla warfare, and WMD including attacks on domestic targets) are issues of special concern and create challenges to religious support planning. Particularly problematic is the growing threat to our increasing information infrastructure through acts of cyber-terrorism.

2-6. Unlike the Cold War era, when threats were measurable and predictable, Army forces today are likely to encounter conditions of greater ambiguity and uncertainty. The U.S. Army must be prepared to respond to worldwide strategic, operational and tactical challenges either alone, or as part of a joint or multinational team. The U.S. Army's UMTs must also be prepared to provide religious support in every contingency or circumstance.

2-7. Mission, enemy, terrain and weather, troops and support available, time available, civilian considerations (METT-TC) includes "civilians" as an operational planning consideration. The UMT requires skills to deal with the religious support requirements of DA civilians, contractors in the area of operations, and indigenous and host nation civilians. With the expected increased urban terrain missions, the UMT will have more opportunity and responsibility to interface with the local people and religious leaders as it supports the command mission.

FORCE PROJECTION

Force Projection Process	
Mobilization	Sustainment
Deployment	Redeployment
Employment	Demobilization

2-8. The Army performs its mission within the context of national security and military strategies. Power projection is the application of the elements of national power—diplomatic, economic, information, and military—anywhere in the world.

2-9. The military aspect of power projection is force projection. Force projection is the deployment of forces to perform missions spanning the whole range of military operations anywhere in the world. Force projection applies to the entire Army, Active and Reserve Components.

2-10. Force projection encompasses a process in which forces are managed, trained, and projected to the theater of operations for any contingency. The force projection process is organized in phases: Mobilization, Deployment, Employment, Sustainment, Redeployment, and Demobilization.

2-11. Credible force projection depends on the capability of the United States to deploy forces rapidly and to perform missions spanning the whole range of military operations. A unit may deploy with short notice, causing tremendous stress on soldiers and systems.

2-12. When the United States decides to employ its military resources, the nation continues to meet the soldier's rights to free exercise of religion by deploying UMTs with the soldiers. During force projection operations, the religious support mission is to provide religious support to the soldiers, authorized civilians, and families of the deployed unit. The environment and type of operation may change, but the religious support mission remains constant throughout the spectrum of operations. The successful provision of religious support to meet requirements of force projection rests upon four pillars (table 2-1).

Table 2-1. Four Pillars of Religious Support

Religious support plan (RSP) {to include the CMRP}	Coordinated with all echelons of command and included in the operation plan/order.
Communication architecture	UMTs must be able to interact with their commands and RS providers at higher, subordinate and adjacent units. (Plan redundancy when possible).
Split-based operations	Concurrent RS provided from projection platform to theater of operations.
Force tailoring	UMTs assigned to operational units and Task Forces to meet unique religious support requirements, liaison with faith-based Non-Governmental Organizations (NGOs) and International Organizations (IOs), and governmental agencies.

LEVELS OF MILITARY ACTIONS

2-13. Levels of military actions include strategic, operational, tactical, and military operations other than war (MOOTW). Military operations in the 21st century will span all levels of military actions, often simultaneously or in rapid sequence. For example, forces on the ground may be engaged in high-intensity conflict one moment, then with changes in the strategic diplomatic or political circumstances, be leading a MOOTW operation the next moment, or vice versa. The Army Chaplaincy of the 21st Century will be organized and equipped to meet the challenges at all levels of actions at any time.

STRATEGIC ENVIRONMENT

"Although U.S. military forces enjoy superiority in many dimensions of armed conflict, the United States is likely to be challenged by adversaries who possess a wide range of capabilities, including asymmetric approaches to warfare, particularly weapons of mass

destruction. The United States cannot predict with a high degree of confidence the identity of the countries and actors that may threaten its interests and security."

--Quadrennial Defense Review Report, 2001

Strategic Religious Support Mission Essential Task List

2-14. The Army's strategy is to achieve and maintain a capabilities-based, threats-adaptive Army that is postured to support our nation's strategy to respond to the full spectrum of crises. To support the Army's strategy, the Chaplaincy must anticipate changes in the strategic environment, technological advances, and evolving Joint war-fighting concepts. The Army's strategic mission essential task list (METL) describes what "well–trained, superbly led, and well-equipped soldiers do for the nation (FM 3-0)." Likewise, the Army Chaplaincy in support provides strategic METL tasks that shape and sustain the spiritual freedoms of the soldier.

Army and Chaplain Corps Mission Essential Task Lists

Army—

- Shape the security environment.
- Respond promptly to crisis.
- Mobilize the Army.
- Conduct forcible entry operations.
- Dominate land operations.
- Provide support to civil authorities.

Chaplain Corps—

- Provide and perform personal delivery of religious support throughout full spectrum of operations.
- Train and equip the Chaplain Corps.
- Advise the commander on matters pertaining to the religious, moral, ethical, and morale climate of the command.
- Advise the commander on matters pertaining to indigenous religions and their impact on military operations.
- Manage mobilization of the Chaplain Corps throughout all Army components.
- Plan and coordinate for religious support with other services, international chaplaincies, and governmental agencies.
- Provide UMTs to implement Combatant Commanders' religious support requirements. (Combatant Commanders are the commanders of the nine unified commands in DOD i.e. USEUCOM, USPACOM, USCENTCOM, USSOUTHCOM, USSOCOM, USTRANSCOM, USJFCOM, USNORTHCOM, USSTRATCOM)
- Provide and perform religious support during military support to civilian authorities (MSCA).

OPERATIONAL ENVIRONMENT

2-15. The operational level of war links the tactical employment of forces to strategic objectives. The focus at this level is on operational art—the use of military forces to achieve strategic goals. Emerging doctrine from all branches is shaping the religious support environment by making increasing requirements for support in this much wider spectrum of operations.

2-16. On the nonlinear, asymmetrical battlefield, with operations characterized by multiple and simultaneous operational missions, religious support tasks become much more challenging. A sound operational religious support plan, executed skillfully, increases the chances of tactical success. It does this by decreasing the tactical UMTs risks and supports the UMT's time management requirements. Operational religious support planning integrates conditions, sequences of support actions, and wise management of resources throughout the theater, to optimize the delivery of religious support.

TACTICAL ENVIRONMENT

2-17. Tactics is the art and science of employing all available means to win battles and engagements. Tactical operations are rapid and dynamic in nature. Specifically, the subject of tactics comprises the actions taken by a commander to arrange units and activities in relation to each other and the enemy. At this level, UMTs are fully engaged with their units maneuvering to the area of operations delivering religious support.

MILITARY OPERATIONS OTHER THAN WAR ENVIRONMENT

2-18. Military Operations Other Than War (MOOTW) includes a wide range of contingencies. It may involve offensive or defensive operations, but most often the primary focus is on stability operations and support operations. During MOOTW, UMTs focus their religious support planning and execution on spiritual care of the soldier over long durations of deployment and serve, when directed, as the commander's liaison with local religious leaders, and faith-based NGOs and IOs.

DECISIVE FULL SPECTRUM OPERATIONS

2-19. Within the operational and tactical levels of military actions, commanders plan and execute military operations. There are four types of operations: Offense, Defensive, Stability and Support. The Army's role differs with each operation (Figure 2-1).

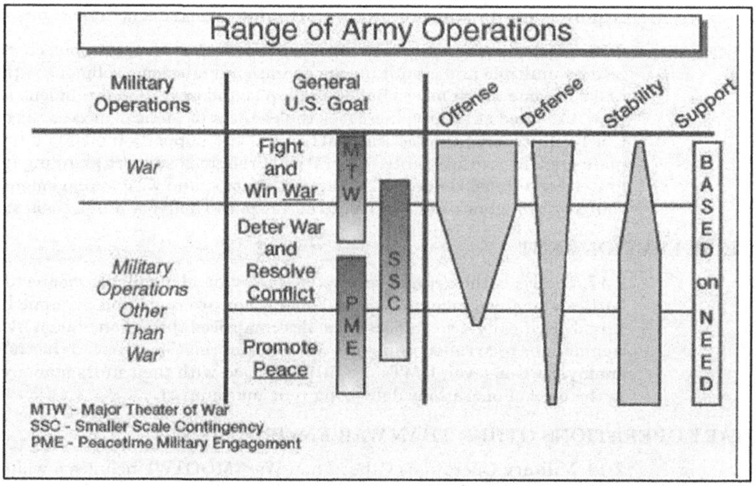

Figure 2-1. Range of Army Operations

OFFENSE

2-20. This is the decisive form of war. Offense is essential to success in all operations—defensive, stability, and support as well as offensive.

DEFENSIVE

2-21. These operations are directed by command to defeat enemy attacks, buy time, economize forces, or develop conditions favorable for the offense.

STABILITY

2-22. Stability operations include a range of actions that Army forces conduct outside the United States and its territories. The main goal is to promote and sustain regional and global stability. Characteristics of stability operations include a diverse and complex environment. These operations are highly demanding on small units and leaders, require intense mental and physical agility, and are often long-term. There are ten types of stability operations:

- Peace operations.
- Foreign Internal Defense.
- Security assistance.
- Humanitarian and civic assistance.
- Support to insurgencies.
- Support to counter drug operations.
- Combating terrorism.

- Noncombatant evacuation operations.
- Arms control.
- Show of force.

SUPPORT

2-23. Support operations are conducted to relieve suffering and help civil authorities prepare for, or respond to, crises. They are divided into two categories: Domestic support operations (DSO) conducted within the United States and its territories, and foreign humanitarian assistance (FHA) conducted outside the United States and its territories. Support Operations usually are conducted to relieve or reduce the results of natural or manmade disasters. They are also conducted to relieve conditions (such as pain, disease, hunger, or privation) that present a serious threat to life or loss of property. There are several forms of support operations.

- ***Relief operations.*** Humanitarian relief focuses on the well being of supported populations. Disaster relief focuses on recovery of critical infrastructure after a natural or manmade disaster.
- ***Support to incidents involving weapons of mass destruction (WMD).*** Assistance is given to civil authorities protecting U.S. territory, population, and infrastructure before an attack by supporting domestic preparedness and protection of critical assets.
- ***Support to Civil Law Enforcement.*** This involves activities related to counter terrorism, counter drug operations, military assistance to civil disturbances, and general support.
- ***Community assistance.*** Provision is given to community assistance for projecting a positive military image, training opportunities, and public relationships.

Religious Support in Peace Operations

2-24. The range of religious support is tailored to support the objectives outlined in the operation plan or order. Unit Ministry Teams must understand each of the four types of military operations in order to plan for, and execute, religious support.

2-25. The Army performs peace operations that range from explicitly peaceful missions to those that have significant applications of combat power. Peace operations are classified as peace making, peace building, and preventative diplomacy. These operations are conducted outside U.S. territories (FM 100-23).

Religious Support in Joint, Combined, and Interagency Operations

2-26. Army units deploy with other military services and agencies of the U.S. government. Increasingly, the U.S. Army operates with other countries as well. Army forces will normally deploy as part of a joint, multinational, and interagency operation. These operations are collectively called unified actions. This is true not only when the military effort is the prime strategic option, but also when other strategic options are preferred.

2-27. Joint operations are operations in which two or more American military departments are employed in the same operation. Joint doctrine is published under the authority of the Chairman of the Joint Chiefs of Staff.

JP 1-05 provides doctrinal guidance for religious support in joint operations. It is important to recognize that Joint publications take precedence over doctrine of the participating service branches.

2-28. Combined operations are operations in which American forces are employed along with forces from other nations. Operational procedures are determined through agreement and coordination by higher headquarters. Provisions for religious support should be included in the operational procedures or policies agreed upon by the participating nations and coalitions; for example, American, British, Canadian, and Australian/New Zealander (ABCA) or the North Atlantic Treaty Organization (NATO).

2-29. Interagency operations are operations in which the military works in tandem with another government agency. The lead agency and organizational relationship are specified in the mission directive. The provision of religious support must be specified for each operation.

2-30. Standardized methods of operating may not be present in many Joint, Combined, or Interagency Operations. Circumstances will occur that cannot be anticipated. UMTs must be flexible and creative in planning religious support for these operations. In each type of operation, the key elements to consider for religious support mission accomplishment are understanding, cooperation, and coordination. These will take place between U.S. chaplains, multinational chaplains, and interagency representatives. This is essential to achieve unity of effort in providing total and comprehensive theater of operation religious support.

Religious Support in Special Operations

2-31. Special Operations Forces (SOF), which consist of specialized Army, Air Force, and Navy units, give the President and the Secretary of Defense additional military options across the full spectrum of operations. These options provide a strategic response capability to prevent future, or stabilize current, international situations with minimum risk to U.S. interests. The theater SOF commander executes special operations (SO) as part of the theater commander's joint special operations effort. Religious support is woven into this specialized Army and joint environment. The religious support planning encompasses all contingencies, whether direct or indirect military actions in support of a Joint Force Command.

2-32. U.S. Army Special Operations Command (USASOC) commands all CONUS-based active and reserve Army Special Operations Forces (ARSOF) units. This command is comprised of five types of units: Special Forces (SF), Rangers, Army Special Operations Aviation (ARSOA), Psychological Operations (PSYOP), and Civil Affairs (CA).

2-33. The command is responsible for readiness and training oversight for all Army Reserve SOF and all outside the continental United States (OCONUS) forward based (deployed) active Army SOF. When directed, USASOC provides mission-ready ARSOF assets, through United States Special Operations Command (USSOCOM), to the regional combatant commanders for employment under their command authority. (See Chapter 5 for religious support tasks.)

Religious Support and Military Support to Civilian Authorities

2-34. Another military environment in which religious support is provided is Military Support to Civilian Authorities (also known as MSCA). These domestic support operations involve the Army in assisting civilian authorities by providing general military assistance. General assistance for domestic support operations includes disaster assistance, environmental assistance, and law enforcement support and community assistance. These operations are conducted within U.S. territories (FM 100-19).

2-35. Religious support in both peace operations and MSCA, as well as Military Assistance to Civil Disturbances (MACDIS), require intentional coordination and cooperation within the Chaplaincy among Army components (active duty, Reserve, and National Guard (NG)), as well as with civil and NGO leaders.

2-36. Civil Support will involve extensive coordination and liaison among interagency, joint, multi-jurisdictional (state and local), and active and Reserve Component (AC/RC) chaplaincies. When state authorities request Federal help, the Federal Emergency Management Agency (FEMA) most often is the lead agency, with DOD supporting. The Secretary of the Army is the executive agent. The Army Chaplaincy works directly for the military command providing RS to soldiers and authorized civilians.

2-37. Reserve Component UMTs provide a unique capability for the Chaplaincy. They have a long-term relationship with state and local religious leaders, institutions, officials, and businesses. They also have special capabilities and are geographically dispersed to conduct religious support operations. UMTs provide and perform the following religious support before, during, and after crises:

- *Planning*. Army National Guard (ARNG) and Reserve component (RC) UMTs are ideally suited to support any locality because of their dispersion throughout the Nation and long-term relationships with local clergy, civil agencies, and institutions. Planning must consist of information management, synchronization with all military and civil assets, equipping and training for all contingency missions, close support of Civil Military Operations tasks, shaping the force religious support structure to meet operational needs and staff integration.
- *Preparing*. Doctrine, organization, training, material, leader development, personnel and facilities (DOTMLPF) issues need to be continuously reviewed by chaplain leadership in all Army components to respond effectively within Civil Support operations. All written guidance and plans need to be continuously updated to meet METT-TC needs and directives. Standard operating procedures (SOPs), UMT battle-books, and other publications need to address all areas to which all UMTs will be directed to respond. Training incorporates many of the collective skills that UMTs already possess and continually update.
- *Responding and executing*. When a crisis occurs, UMTs of selected units, in support of the lead agency, will provide and perform direct religious support to military units, will perform liaison responsibilities with State and Federal agencies pertaining to religious issues, will conduct religious risk assessment for the command, and will liaison with civilian religious leaders.

- *Follow-up*. In the days following a crisis, UMTs provide and perform religious support directly to military forces as they are employed after a hostile event, to include the following tasks: mass casualty religious support, critical event debriefings, liaison with local religious leaders, etc.

2-38. The Medical Command (MEDCOM) plays a key role in response efforts. Existing MEDCOM units may be tailored to meet the METT-TC requirements. MEDCOM UMTs will bring specific skills in mass casualty, trauma, crisis intervention, and critical event debriefing ministry to these events. RS planning must be well integrated within this health care response. Army AC/RC deployable hospitals may move forward to the area of operations. Military medical facilities may be used as a trauma center when that capability does not exist locally, or when called for under National Disaster Medical System (NDMS). Large numbers of casualties resulting from WMD will challenge the UMT and other caregivers, resulting in increased need for critical event debriefing and other types of intervention ministries. Special Medical Augmentation Response Teams–Pastoral Care (SMART-PC) play a unique religious support role in response ministry. SMART-PCs, as well as the chaplain's role in support of Special Medical Augmentation Response Teams–Stress Management (SMART-SM), are discussed in depth in chapter 6 of this FM.

SUMMARY

2-39. Religious support provided during a military operation can occur under widely differing circumstances. The UMT's basic mission to provide religious support to the soldier, however, remains constant. What differs is the operational environment in which the religious support is provided. The differing types of military operations determine the specific nature of religious support.

2-40. The U.S. Army's UMTs must be prepared to provide religious support in every contingency or circumstance. With the expected increase in urban terrain missions, the UMT will have more opportunity and responsibility to interact with the local people and religious leaders in supporting the command mission. Additionally, METT-TC includes "civilians", which include both authorized DA Civilians, civilian contractors on the battlefield, as well as indigenous and host nation civilians. These civilians are included as an operational planning consideration for the UMTs to determine who, what, when, and where to provide and perform religious support.

2-41. In all types of operations across the full spectrum, key elements to consider for the religious support mission accomplishment are: understanding, cooperation, and coordination between U.S. chaplains, multinational chaplains, and interagency representatives. This is essential to achieve unity of effort in providing total and comprehensive theater of operation religious support.

2-42. There are no boundaries between strategic, operational, tactical, and MOOTW environment. Over time in the battlespace of a theater, operational activities link the tactical employment of forces to strategic objectives. The Army strategic METL describes what "well –trained, superbly led, and well-equipped soldiers do for the nation (FM 3-0)." Likewise, the Army

Chaplain Corps, in support, provides a strategic METL that helps shape and sustain the spiritual center of the soldier.

2-43. Civil Support will involve extensive coordination and liaison among interagency, joint, multi-jurisdictional (state and local), and active and reserve component (AC/RC) Chaplaincies. The RC UMT provides a unique capability to the Chaplaincy. They have long-term relationships with state and local religious leaders, institutions, officials, and businesses. Also, RC UMTs have special capabilities and geographic dispersion to conduct religious support operations. Religious Support tasks will be in concert with the Army's role in Civil Support which follow in these categories: force protection, support to crisis management, support to consequence management, protection of critical assets, support to counter-terrorism, deterrence/defense against strategic attack, and MSCA missions.

RESOURCES

2-44. The following is a list of main resources utilized for writing and updating this doctrine.

- AR 165-1.
- JP 1-05.
- JP 5-00.2.
- JP 3-0.
- JP 3-05.
- FM 3-0 (100-5).
- FM 16-1.
- FM 22-100.
- FM 1 (100-1).
- FM 100-19.
- National Military Strategy
- Joint Vision 2010
- Army Vision 2010
- A National Security Strategy For a New Century, May 1997
- U.S. Army Chaplaincy Strategic Plan FY 2000-FY 2005
- Chief of Chaplains MSCA Concept Paper, September 2000
- Chaplain Support Force XXI White Paper
- Quadrennial Defense Review Report, September 2001.

Chapter 3

UNIT MINISTRY TEAM OPERATIONS

"I was determined that whatever else might happen next time [post-W.W.I], at least I myself would be prepared, and trained, and ready when the time came. I had at times a kind of inward feeling that the call would come, to me personally, and in my prayers morning and evening I used to ask that I might be given help and strength so that I might not fail when put to the test."

--Field Marshall Montgomery, The Path of Leadership

"Everything is simple in war, but the simplest thing is difficult."

--Carl Von Clausewitz, On War

This chapter establishes the essential elements of Religious Support (RS) planning, coordination, and execution efforts in all phases of operations. It examines the following key points:

- General Religious Support planning, preparation, coordination and execution overview.
- Religious Support Planning for Force Projection.
- General Responsibilities of the UMT During Religious Support Operations.

CONTENTS

GENERAL RELIGIOUS SUPPORT PLANNING, PREPARATION, COORDINATION AND EXECUTION OVERVIEW

3-1. To meet the religious needs of soldiers and other authorized personnel in all operations, the Unit Ministry Team (UMT) prepares a comprehensive religious support plan that is staffed and integrated into every operational phase. The UMT advises the commander on the plan for accomplishing the religious support mission. The plan encompasses RS required from training and mobilization, to redeployment and demobilization.

3-2. The commander is responsible for the execution of the Religious Support Plan (RSP). The Chaplain, as a member of the commander's personal staff, is responsible to the commander to plan and implement the following:

- Developing the RSP.
- Executing the approved RSP.
- Use of available religious support resources.
- Allocation of assets.
- Location of UMTs.
- Coordination with indigenous religious leaders.
- Coordination with religious NGOs and IOs in coordination with Civil Military Operations leaders.
- Religious support for all authorized civilians in the AO.

PLANNING

3-3. The UMT participates in the collaborative planning process with the other staff sections known as the Military Decision-Making Process (MDMP). Religious support planning is continuous, detailed, and systematic. It examines all factors relating to the religious support of an operation. The planning process builds on the religious support section of the unit's tactical standard operating procedures (TSOP). It becomes specific to an operation through the preparation of a religious support estimate (RSE). The product of this planning process is a religious support plan (RSP). The RSP is included in the operations plan or operations order (OPLAN/OPORD). Inclusion may be in the plan/order as an appendix, annex or TAB. The specific placement of the RSP is dependent upon the particular unit's protocol for constructing operation plans and orders. The RSP informs the command and staff of the concept of religious support, and guides the work of the UMT. When appropriate, the RSP can be contained in a matrix format. A model for such a RSP matrix is contained in Appendix D of this manual. The use of a RSP matrix is appropriate when UMTs are assigned or attached to units that publish their OPLANS/OPORDS in a matrix format. The RSP matrix model is a self-explanatory form that clearly and concisely articulates how the UMT will provide religious support during the operation. The use of a RSP matrix is highly recommended when time constraints impact the planning process and the UMT needs a ready tool to direct and encapsulate its religious support planning.

KEY PLANNING CONSIDERATIONS:
- Commander's Intent
- Concept of Operations
- Concept of Logistics
- Concept of Medical Support
- ROE
- METT-TC of the battlespace/theater of operations

Religious Support Estimate

3-4. The estimate process is a problem solving procedure to assist the commander in determining the best course of action (COA) to accomplish a mission or task. Staff estimates are designed to supplement the commander's estimate within the respective areas of expertise. A staff estimate consists of significant facts, events, conclusions and recommendations on how best to use available resources to accomplish the mission. Upon receipt of the commander's guidance, each staff section develops its own staff estimate relating how its area of interest will affect the mission.

3-5. The chaplain develops the Religious Support Estimate (RSE) that becomes the basis for the Religious Support Plan (RSP). Development of the RSE ensures the careful and intentional planning of religious support. In the development of the RSE, the chaplain considers the —

- Location and movement of the UMT during all phases of the operation.
- Logistical and communication support required for specific missions.
- Replacement and/or augmentation of UMT members in the event UMT members become casualties or are unable to meet all RS needs during especially intense fighting.
- Religious, moral, and morale issues that affect the soldiers and the unit.
- Impact of indigenous religions on the military operation.

BASIC PRINCIPLES OF RS PLANNING AND COORDINATION

- Plan early and continuously.
- Use guidance from higher HQ.
- Exploit all available RS assets.
- Use the lowest echelon capable of furnishing effective RS.
- Use the most effective RS means.
- Furnish the type of RS requested.
- Avoid unnecessary duplication.
- Coordinate RS along command and technical channels.
- Provide adequate support.
- Provide rapid coordination.
- Provide safeguards and survivability.
- Establish RS coordination measures.
- Establish communications support.

3-6. The preparation of the estimate follows a logical, continuous, and methodical process to provide the appropriate religious support for the commander's selected COA. In conjunction with the other staff sections, the UMT prepares the RSE using the following steps:

1. Gather facts and assumptions.
2. Analyze the mission.
3. Develop courses of action.
4. Analyze and compare courses of action.
5. Make a recommendation.

Step 1: Gather Facts and Assumptions

3-7. The staff gathers information concerning the assigned mission and divides the information into two categories: facts and assumptions. Facts are statements of known data concerning the situation, and include the following:

- Enemy and friendly dispositions.
- Available troops.
- Unit strengths.
- Materiel readiness.
- Mission and the commander's intent.

3-8. Special religious needs of soldiers in the unit are key facts for the RSE. The team develops a Religious Preference Profile (RPP) using data from a Standard Installation/Division Personnel System (SIDPERS) ad hoc query. (See AR 680-29 for religious preference codes.) The report is a statistical analysis of the unit's religious demographics and provides the stated religious preference of soldiers assigned to that unit.

3-9. When facts essential to planning for RS are not available, the UMT must make assumptions. An assumption is a supposition about the current or future situation, assumed to be true, though without positive proof. In the absence of facts, assumptions are used to fill gaps. Assumptions are necessary only when they bear directly upon the planning or the execution of the plan. As new information becomes available, some assumptions become facts and some are modified or discarded.

3-10. Mission, Enemy, Terrain and Weather, Troops and support available, Time available, and Civil Considerations, or METT-TC, is a useful outline by which to consider and formulate facts and assumptions essential for developing an RSE.

Mission, Enemy, Terrain and Weather, Troops and support available, Time available, and Civil Considerations (METT-TC)
Mission: The UMT identifies specified or implied religious support tasks from the higher headquarters' OPLAN or OPORD. For example, a requirement to provide area religious support may be implied but not specified by a higher headquarters' task organization. Other considerations may include protection of religious shrines, artifacts or sacred places in the AO.
Enemy: In this part of mission analysis, the team identifies how the enemy may threaten the religious support mission. For example, enemy activity in rear areas may restrict team travel and necessitate additional area support requirements. The team may also consider how religion plays a part in the enemy's capacity or inclination to fight.
Terrain and Weather: The UMT assesses the terrain and weather in the AO. Traffic-ability, weather extremes, and the amount of daylight are factors that will affect the religious support mission. (Consider observation and fields of fire, cover and concealment, obstacles, key terrain, and avenues of approach (OCOKA).
Troops: Religious support is focused upon soldiers. What soldiers are doing often determines their availability for religious support. The UMT adjusts its plan to match the soldiers' schedules to meet their needs. The UMT must also assess the level of anxiety and fatigue among the troops to shape its ministry to the soldiers more effectively. The UMT will also advise the commander on the morale, moral, and spiritual health and welfare of soldiers.
Time Available: The UMT assesses the time required for planning and the time available for executing the religious support mission. The amount of planning time determines the detail of the plan. The team also considers the travel time to each unit, preparation time, and available light.
Civil Considerations: The UMT researches and advises the command on the religious attitudes and activities of the civilian population, religious and cultural dynamics, religious organizations, and religious leaders within the battlespace. Coordination with CMO personnel is essential. The chaplain advises the commander regarding civil concerns and their impact on the mission: • Understanding the operational religious environment. • Recognize religious implications for courses of action during planning. • Anticipate and control negative impacts on the local religious climate. • Consider long-term effects of operations on the indigenous religious climate. • Ensure legitimacy and credibility of friendly forces. • Balance operational priorities with concern for religious freedom. • Minimize negative effects of operation on unit soldiers belonging to the same faith group as the local population.

Step 2: Analyze the Mission

3-11. Mission analysis begins with a review of the intent of commanders one and two levels higher. From this review, the commander and staff derive the specified tasks that the unit must perform to accomplish the mission. The unit must also identify implied tasks and determine which tasks are essential. A tentative list of essential tasks is presented to the commander for approval. Based upon the approved list, the staff presents the commander with a restated mission for approval. Commanders disseminate the restated mission as a warning order.

3-12. The UMT participates with the rest of the staff in mission analysis by identifying specified and implied religious support tasks. The UMT also considers any religious factors that may directly affect the mission. (FM 101-5, Appendix J)

3-13. During mission analysis, in addition to the RSE, the UMT brings expertise to the staff team in studying the following areas:

- ***Unit Morale.*** The UMT should have a sense of the morale of the unit. Factors such as whether the troops are fatigued, are excited or dreading the mission, and the spiritual health and welfare of the unit all impact on unit morale.
- ***Ethical and moral implications.*** The chaplain can assist the staff in considering the ethical and moral implications of a particular plan as it impacts on the lives of leaders, soldiers, and the local civilian population. For example, this may include compassionate treatment of EPWs, detainees, and displaced persons.
- ***Impact of local religion on the mission.*** The chaplain can assist in the religious assessment of an AO. The UMT should familiarize themselves with the local religions and their significance within the area of operations. What and where are the places that are considered sacred? What are the significant religious holidays and how are they celebrated? How could this impact the operation or the timing of the operation? (See Appendix F).

Step 3: Develop Courses of Action

3-14. Course of action (COA) development is based on mission analysis and on the facts and assumptions developed earlier. After receiving the commander's guidance, the Chief of Staff (CofS) or the executive officer (XO) leads the staff in developing COAs for accomplishing the mission. As the staff begins to develop COAs, the chaplain considers the religious and moral/ethical implications of each proposed COA and how to provide religious support. While the operations officer formulates the tactical possibilities, the chaplain and other staff officers consider how to integrate their functional areas into each COA. In the midst of COA development, the use of Chaplain Detachments (CD) by those units to which they have a doctrinal association must be considered and implemented into the COAs and the eventual RSP. Likewise, units with a doctrinal association with CDs must proactively request CDs through appropriate staff actions to ensure their operational employment.

Step 4: Analyze and Compare Courses of Action

3-15. After the staff develops several COAs, it analyzes each one. COA analysis consists of two parts: war-gaming and COA comparison. Through this process, the staff identifies the best COA to recommend to the commander.

3-16. Note that the chaplain may have a different kind of input depending on the kind of mission that is being planned. For units preparing for offensive or defensive actions, often the chaplain's role is more involved with integrating with the staff and being a team player while developing the RS aspect of the plan. In a stability and support operation, a chaplain may have more to contribute in the plan development process. In both cases, the chaplain will want to anticipate the COA toward which the commander may be leaning.

3-17. Additionally, the chaplain will want to pay particular attention to the casualty flow estimates and concepts of support. During the war gaming phase of the MDMP, chaplains must pay particular attention to where the casualties are expected, remembering that this critical ministry is one of the three major functions of comprehensive religious support. If the chaplain participates in the war gaming phase of the analysis, the chaplain will have a much clearer understanding of the scope of the mission when the COA is finally decided upon and then rehearsed.

Step 5: Make a Recommendation

3-18. After choosing the best COA, the staff briefs the commander. The commander may agree with the staff's recommendation, modify it, or select another. Once the commander decides on a COA, the UMT uses the concepts of support developed during the RSE process to prepare the Religious Support Plan (RSP).

3-19. The commander and staff begin preparations for the mission rehearsals. The chaplain needs to participate in the rehearsals in order to test the effectiveness of the RSP. Afterwards an assessment is made as to whether the plan actually meets the needs of the troops according to the commander's guidance.

Note: The adjutant (S1), supply officer (S4), and the chaplain are often not included in the accelerated MDMP process. However, this does not preclude the UMT from preparing RSEs and RSPs. Staff coordination and proper analysis is always necessary.

Religious Support Plan

3-20. Based on the selected COA, the UMT completes the details of its plan for inclusion in the OPLAN/OPORD. The religious support plan (RSP) assigns religious support responsibilities, defines area support requirements, and authorizes coordination between subordinate units.

3-21. Additional information found in the RSP may include the following:

- Concept of support for special faith groups and special holy days.
- Concepts of support for joint and multinational religious services support.

- Religious support requirements of all authorized civilians.
- Religious support for EPWs and Detainees in accordance with the Geneva Convention.
- Command guidance regarding liaison responsibilities with faith-based NGO/IOs and liaison support, in coordination with CMO personnel, to religious leaders in the AO.

3-22. Priority of religious support is—

- Unit military personnel.
- Other military personnel in AO.
- Authorized civilians such as DOD employees and contractors.
- EPWs and Detainees.

Planning Religious Support in Joint Operations

3-23. Just as in the Army planning process, the UMT is included in the system of joint planning. The Religious Ministry Support Team (RMST) (chaplain and chaplain assistants as designated under JP 1-05, which the Army designates as UMT) at each echelon is intricately involved with the orders process. Planning religious support operations must encompass and consider—

- Denominational assets.
- Command relationships.
- Location of RMSTs.
- Logistical requirements.

(See JP 1-05 for more detail.)

Planning Religious Support in Multinational Operations

3-24. Religious support planning with multinational forces in an operation must include the differences in religious support doctrine (or military custom of providing religious support), capabilities, force structure, logistics, and culture (denominational representation, language, supervision, etc). Communication, cooperation, and coordination are essential in planning multinational religious support operations. Senior force RMSTs need to understand the capabilities and limitations of religious support with each country represented within the coalition. Planning must remain flexible due to these differences while emphasizing the similarities that may include, language common between countries, denominational backgrounds of chaplains, and locations for area support. Senior U.S. RMSTs must be highly sensitive to allocating religious support assets of another country and should only do so with prior consultation and coordination. Planning considerations include—

- Communication requirements and limitations.
- Denominational representation.
- Language difficulties.
- Limitations in formal chaplain structures.

Rules of Engagement

3-25. Rules of engagement (ROE) specify the circumstances and limitations in which forces may engage the "enemy". Many factors influence ROE, including national command policy, mission, operational environment, commander's intent and the law of land warfare. Properly written ROE are clear and tailored to the situation.

3-26. Chaplains at all echelons help commanders apply the ROE by advising on the moral/ethical implications of proposed COAs. Senior-level chaplains consider the restrictions and constraints of the ROE when recommending religious support policy to the commander.

PREPARATION

3-27. Preparation is the link between planning and execution. While the commander and staff are planning, the unit prepares for the operation based on a warning order issued earlier by the commander. Once the plan is complete, preparations unique to the mission may begin. Upon completion of the individual and team preparations, the UMT provides religious support to soldiers while they complete their preparation for the operation.

Rehearsals

3-28. Units conduct rehearsals during preparation for combat. In addition to maneuver rehearsals, units conduct combat service support (CSS) and casualty evacuation rehearsals. Rehearsals provide an opportunity to refine the RSP. The UMT must be present at the rehearsal, and integrate the RSP into it, to fully understand the scheme of maneuver, support, and other critical pieces of information shared. If time permits, UMTs may conduct their own RS rehearsals following those of the unit. Such RS rehearsals offer UMTs an opportunity to resolve communication issues, as well as, build a comprehensive understanding of each UMT's religious support mission.

Individual and Team Preparation

3-29. The chaplain and chaplain assistant prepare themselves for combat like other soldiers of the unit. This preparation must be physical, mental and spiritual. The chaplain and chaplain assistant must also insure that their families are ready for coping with the challenges of deployment.

3-30. Team preparation includes load plans, pre-combat checks, updating information, map reconnaissance, travel planning, rehearsals, and mission coordination. Other preparations include securing equipment for the operation such as worship aids, devotional literature, music, and liturgical items.

Pre-Combat Inspections and Preparation

3-31. The UMT conducts a pre-combat inspection according to the unit TSOP. This inspection ensures the UMT possesses sufficient supplies and equipment to successfully execute the operation. On every mission, the team must carry food, water, and other survival equipment. The chaplain assistant prepares and updates the team's load plan and inventories and arranges the necessary supplies and equipment for movement.

Information Update

3-32. The UMT updates mission information regularly— especially before departing from its operational base. The tactical situation can change rapidly on the battlefield. The team checks the situation map, situation reports (SITREPS), logistics reports, and the latest fragmentary orders (FRAGOs) or warning orders. If there are significant changes in the situation, the team modifies its plan accordingly.

Training

"The key to winning on the battlefield is the understanding of "how we fight" and the demonstrated confidence, competence, and initiative of our soldiers and leaders. Training is the means to achieve the tactical and technical proficiency that soldiers, leaders, and units must have to enable them to accomplish their missions."

FM 7-0, TRAINING THE FORCE

3-33. Training is the cornerstone of preparation for military operations. The Army's primary activity during peacetime is realistic, battle-focused training oriented on the unit's mission essential task list (METL). Installation chaplains, through their training managers, and in coordination with subordinate UMTs and unit operations/training officers (S3/G3), ensure that UMTs have adequate technical and tactical training for deployment and combat. The training objective is operational readiness—units and soldiers that are tactically and technically proficient.

3-34. Tactical proficiency is a demonstrated understanding of the Army's war fighting doctrine and tactics. To provide effective religious support in combat, the UMT must know its unit's standard operational procedures, tactics, and techniques. The UMT must also maintain the requisite skills of field craft essential for survival on the battlefield. Tactical training is primarily the responsibility of the unit to which the UMT is assigned or attached.

3-35. Technical proficiency is the demonstrated ability to perform all tasks related to one's functional area. Technical proficiency is required for both garrison and field missions. Technical proficiency is acquired from schools, self-study and training with other UMTs in performing the unit's and installation's religious support mission. Training and continuing education conferences, functional courses offered by the U.S. Army Chaplain Center and School (USACHCS), and courses offered through the Officer and NCO Education System (OES and NCOES) supplement the installation religious support-training program.

"To understand firepower and its application you have to use it, actually practice with it and see it work. This means maneuvers in the field, maneuvers that reproduce as realistically as possible, within the bounds of prudence, the actual conditions a soldier will encounter on the battlefield. Only so can you condition, not only the human body, but also the human spirit, to face and survive the stresses and strains the soldier will encounter in battle. And in future battle, if it comes, these stresses and strains will be immeasurably greater."

General Matthew Ridgeway

3-36. The UMT must plan for and participate in religious support training for both combat and garrison environments. Furthermore, the UMT must participate fully in both the unit tactical and religious support training to assure their tactical and technical proficiency.

Assessment of Unit Ministry Team Readiness

3-37. As the commander's personal staff officer for religious support, chaplains at brigade and higher levels are responsible for assessing the tactical and technical proficiency of UMTs in subordinate units. They conduct assessments of the UMT during field training exercises (FTXs), external Army Training and Evaluation Program (ARTEP) events, unit training and Command Inspections. Observer controllers provide additional assessment assistance when the UMT is involved in Battle Command Training Program (BCTP) events or Combat Training Center (CTC) rotations. These CTC assessments for UMTs are important for readiness analysis and training.

3-38. Chaplains and chaplain assistants participate in the installation's Command Master Religious Plan (CMRP) to help soldiers develop the personal spiritual readiness needed to sustain them during combat. Spiritual readiness is also essential to sustain soldier families while the unit is deployed. As part of the installation's CMRP, the UMTs conduct worship services, offer religious education, and provide spiritual fitness training and other activities in installation chapels and facilities, and in the theater of operations. Spiritual readiness applies to all levels of the command. Planning for spiritual readiness programs must be part of the preparation for combat for each soldier and authorized personnel.

Religious Support Preparation in Joint Operations

3-39. Coordination between all RMSTs is critical for full implementation of the RSP. Total synchronization between all religious support elements is key to a comprehensive plan. The joint task force (JTF) RMST considers the following additional tasks for operational preparation:

- Identifying RMST logistical requirements.
- Personnel staffing and augmentation requirements.
- RSP integration within the operational plan.
- RSP rehearsals.
- Spiritual Nurture programs for the command and staff.
- Loading of equipment and material.

3-40. The Command JTF RMST works closely with the Joint staff for RMST closure within the Joint Operational Area (JOA), ensuring communication and synchronization of the RSP.

Religious Support Preparation in Multinational Operations

3-41. Each nation brings different capabilities and limitations to the Combined-Joint Task Force (CJTF). Task organizing along functional lines vice national lines can help fill capability gaps, and designs each nation's role so they are full participants. Comprehensive religious support information disseminated to all participating nations describing religious support operational expectations will prepare all combined participants for spiritual and religious coverage. UMTs must logistically prepare themselves to meet the many diverse and varied religious and cultural differences. Communication limitations and differences must be prepared for such as language, terminology, systems, and distance. Liaison and training can offset some of the challenges.

"A chaplain visits our company. In a tired voice, he prays for the strength of our arms and for the souls of the men who are to die. We do not consider his denomination. Helmets come off. Catholics, Jews, and Protestants bow their heads and finger their weapons. It is front-line religion: God and the Garand."

Audie Murphy, To Hell and Back

COORDINATION

3-42. Coordination consists of three categories: mission, internal, and external.

Mission Coordination

3-43. The size of the AO and unit dispersion challenge the UMT's ability to coordinate for religious support. Communications equipment is critical for timely and effective coordination. Also, key communications systems allow the UMT to be available for emergency tasks. Early planning and coordination will help to eliminate potential problems. Coordination is both internal to the participating units and UMTs, and with external units, agencies, and UMTs.

Internal Coordination

3-44. As a personal staff officer, the chaplain works for the commander, advises the commander, and executes the commander's religious support program on his behalf. The UMT coordinates the execution of the commander's religious support program with the Chief of Staff or XO. The team plans cooperatively with all staff sections for the exchange of information.

External Coordination

3-45. Because of the rapidly changing dynamics, conditions, and circumstances on the modern battlefield, coordination for area support is difficult. Although religious support execution is decentralized, UMTs must synchronize religious support within their subordinate units. Establishing synchronization procedures with subordinate commands before deployment reduces the communication and coordination challenge for the UMT.

3-46. Military communication channels streamline information distribution on the battlefield. The team uses command, staff, and technical channels for communication, guidance, and coordination.

- ***Command:*** The command channel is the official communication line between headquarters for passing orders, instructions, and tasks to subordinate units. Command channels link commander to commander. Within the authority granted them, staff officers use command channels when acting for the commander. It is seldom if ever used by the UMT.
- ***Staff:*** The staff channel is the staff-to-staff link between headquarters. It is the primary channel the UMT uses for planning and coordinating religious support. Through this channel, the team transmits and coordinates planning information and provides staff supervision of religious support in subordinate units.
- ***Technical:*** The technical channel is the link between members of the same branch or discipline to convey technical instructions and guidance. Chaplains use the technical channel to communicate with UMTs of other

units about the technical aspects and details of religious support. It is not used to send official reports, to request religious support, or to send routine information between teams. This information is communicated through command or staff channels.

EXECUTION

3-47. The execution of the religious support mission on the battlefield is decentralized and performed by UMTs in accordance with the RSP. Adjustments to the plan are coordinated between affected units.

RELIGIOUS SUPPORT PLANNING FOR FORCE PROJECTION

MOBILIZATION

3-48. United States military strategy rests upon the ability to rapidly respond to crises. Responding to crises requires the Army to deploy its forces and may require mobilization of Reserve Component (RC) resources. Deployment comprises those activities required to prepare and move the force and its sustainment equipment and supplies to the AO.

Religious Support during Mobilization

3-49. Mobilization is the activation of the U.S. Army Reserve (USAR) and federalization of the Army National Guard. Mobilization provides commanders with RC forces, manpower, facilities, and logistics. FM 100-17-3, Reception, Staging, Onward Movement, and Integration (RSOI), provides additional information on RSOI procedures. It expands the Army's capability to respond to crises. Mobilization is conducted in five phases:

- Phase I. Planning.
- Phase II. Alert.
- Phase III. Home Station.
- Phase IV. Mobilization Station.
- Phase V. Port of Embarkation.

Mobilization Phase I: Planning

3-50. Planning begins long before a unit is mobilized. During peacetime, Active component (AC) and Reserve component (RC) UMTs coordinate with AC units for their mobilization planning. This planning takes into account operational and training requirements, equipment status, and readiness. It also includes the impact of mobilization on soldiers, families, and communities.

3-51. The primary planning tool for mobilization is the Unit Ministry Team Mobilization Systems Planbook (MOB Planbook). The MOB Planbook integrates all phases and levels of mobilization, and it provides a detailed guide for integrating mobilization planning.

3-52. Installation, Continental United States Army (CONUSA), STARC, and USARC chaplains are key to Reserve Component UMT mobilization. Each maintains a copy of the UMT MOB Planbook. Organizational MOB Planbooks contain the same basic information, but each is tailored for the particular headquarters. The MOB Planbook specifies what must be accomplished during each phase of mobilization. While many of the tasks in the

MOB Planbooks are unique, they are designed to integrate the mobilization process.

3-53. A unit's success is directly related to the quality of its preparation before it conducts operations. A representative listing of the mobilization planning responsibilities for installation, CONUSA, STARC, and USARC chaplains is detailed in Chapter 7. Planning considerations—

- Does the home station require special planning or preparation?
- Where is the home station in relation to support resources?
- How does the UMT provide family support briefings?
- How will the UMT provide required religious support if some parts of the unit remain at home station for an extended period of time, while others elements are sent forward to the mobilization station (MS)?
- Does the UMT go forward with the advance party or with the main body of the unit?
- How do durations in personnel holding areas, movement, or separation influence soldier morale?
- What religious support activities would address morale problems in the unit?
- What actions are required to close out civilian responsibilities?
- What coordination is required with faith groups for religious coverage during mobilization?
- What supplies are required for religious support?
- What religious support is required as the unit transitions to war?
- What training is required to bring the UMT to validated status for deployment?

3-54. Thorough planning and realistic training before a mobilization reduces confusion between all key UMT participants. Mobilization training gives both AC and RC UMTs an opportunity to practice their mobilization religious support missions, and an opportunity to assess the adequacy of mobilization plans. Installations assist with the following RC UMT training events:

- Annual training (AT).
- Individual Mobilization Augmentee (IMA) training.
- Individual training events, Individual Ready Reserve (IRR), and Chaplain Candidate training.
- AC/RC Sustainment Training.
- Training for UMTs in Garrison Support Units (GSUs).
- AC / RC Annual Chaplaincy Regional Sustainment Training (CREST) as mandated by the Chief of Chaplains.

Planning Considerations:

- Does the home station require special planning or preparation?
- Where is the home station in relation to support resources?
- How does the UMT provide family support briefings?
- How will the UMT provide required religious support if some parts of the unit remain at home station for an extended period of time, while others elements are sent forward to the mobilization station (MS)?
- Does the UMT go forward with the advance party or with the main body of the unit?
- How do durations in personnel holding areas, movement, or separation influence soldier morale?
- What religious support activities would address morale problems in the unit?
- What actions are required to close out civilian responsibilities?
- What coordination is required with faith groups for religious coverage during mobilization?
- What supplies are required for religious support?
- What religious support is required as the unit transitions to war?
- What training is required to bring the UMT to validated status for deployment?

Mobilization Phase II: Alert

3-55. The Alert Phase begins when the unit receives the alert order of a pending mobilization order. The UMT reports immediately to its home station and the unit conducts the final screening of soldiers and cross leveling of personnel and equipment. The Alert Phase ends on the effective date of the unit's entry on active duty.

Note: RC chaplains cannot be mobilized (i.e. cannot be brought onto Active Duty) until they have completed all phases of the Chaplain Officer Basic Course (CHOBC). AC chaplains are non-deployable until the sustainment phase of CHOBC is completed. (See AR 165-1)

Mobilization Phase III: Home Station

3-56. Phase III begins on the effective date of the unit's mobilization. The effective date of mobilization is when RC chaplains and chaplain assistants come on Active Duty. During this phase the unit begins its transition to active duty. At home station the unit inventories property, dispatches an advance party, and moves to the mobilization station (MS). This phase ends when the unit arrives at the MS.

3-57. UMTs must balance the time spent in their own deployment preparation with providing religious support to soldiers and families.

Mobilization Phase IV: Mobilization Station (MS)

3-58. Phase IV begins when the unit arrives at the mobilization station. Activities at the MS include processing personnel and equipment, accessioning the unit into the active structure, cross-leveling, and soldier readiness processing. The UMT, along with the unit, engages in individual and collective training and validation exercises for survival on the battlefield. The goal of the unit during this phase is to achieve mission capability in the shortest possible time. The MS phase ends when the unit arrives at the port of embarkation (POE).

Note: The unit's line of authority shifts during this phase from the CONUSA to the MS commander.

Mobilization Phase V: Port of Embarkation (POE)

3-59. This phase begins with the arrival of the unit at its POE. Actions at the POE include preparing and loading equipment and manifesting and loading personnel. The POE phase ends with departure of personnel and equipment from the POE.

UNIT RELIGIOUS SUPPORT DURING DEPLOYMENT

3-60. Deployment is the planning, preparation and movement of forces and support base from any location to an area of operations in response to military need or crisis. Deployment may take the form of a forcible or permissive entry. Deployments are conducted in five phases:

- Phase I. Pre-deployment Activities.
- Phase II. Movement to Ports of Embarkation.
- Phase III. Strategic Lift.
- Phase IV. Theater Base Reception.
- Phase V. Theater Onward Movement.

Deployment Phase I: Pre-deployment Activities

3-61. Pre-deployment activities begin when a unit receives a deployment notification. These activities include mission briefings and planning, family pre-deployment briefings, equipment loading, chapel closeout, and the transfer of property accountability.

Mission Planning

3-62. As discussed above, the commander and staff begin the MDMP after they receive the deployment notification and they review contingency plans for completeness and accuracy. The UMT, along with the rest of the staff, begins to prepare an estimate of the situation. The team balances planning and preparation with providing religious support.

3-63. Religious support planners at higher echelons integrate the broader scheme of the operation into the planning process. They assess lines of authority, task organization, and faith group needs to determine the required number and mix of UMTs. Senior planners monitor the tailoring of the deploying force. When a JTF or Army Forces (ARFOR) commander forms an ad hoc headquarters, senior chaplains must monitor its composition to ensure the inclusion of a senior chaplain on the staff.

3-64. If the theater chaplain is not scheduled to deploy immediately, he designates a senior chaplain to be responsible for coordinating religious support in theater during each stage of the deployment preceding his arrival.

Chapel Transitions

3-65. The installation chaplain is responsible for all religious support facilities on the installation. Deploying UMTs coordinate chapel transition with the installation chaplain who decides whether to continue, combine, or suspend specific chapel programs and services. The UMTs transfer property accountability to the installation chaplain before departure. Deploying UMTs, in concert with non-deploying UMTs, provide transition activities for soldiers and their families.

Load Lists and Plans

3-66. The UMT must determine what to ship and what to carry. Some religious support supplies (communion bread, wine or grape juice) are consumable and cannot be pre-positioned with other war stocks. Sacred scriptures, religious literature, rosaries, crosses and medals can be pre-positioned. Equipment and supplies may not be available in theater for an unforeseen length of time. The UMT must include in the TSOP a religious support supply load list and plan.

Family Support Activities

3-67. As information about the deployment becomes available, the unit, as directed by the commander provides pre-deployment briefings to families (typically the S1 coordinates such briefings). The information given will be constrained by operational security requirements. Pre-deployment briefings include information on services that will be available on the installation during the deployment. Information will also be discussed regarding the advantages or disadvantages of family members remaining in the military community or returning to their community of origin. UMTs participate in pre-deployment briefings to provide religious support information to families.

3-68. Religious support provided to soldiers and their families directly contributes to the success of mobilization and deployment. Religious support fosters family wellness, a substantial factor in soldier readiness. Lessons learned from previous deployments and other research establish that soldiers deploying while anxious about personal and family problems are more vulnerable to combat stress. They are more likely to experience panic, poor judgment, battle fatigue, and the loss of the will to fight.

3-69. A lack of sufficient information, concern for coping without a spouse or parent, and the trauma brought on by a soldier's departure can generate stress within the family. Because families need religious support during this period, many will turn to the UMT. The deploying team must balance the needs of soldiers and families with its own preparation for departure. The non-deploying UMTs must assume much of the religious support responsibility for the families of the deploying soldiers.

3-70. It must also be mentioned that while care for the families of fellow soldiers is a responsibility for the UMT, the families of UMTs will also need care and religious support. Non-deploying UMTs can assist these families through the deployment process.

Deployment Phase II: Movement to the Port of Embarkation

3-71. When pre-deployment activities are concluded, the unit begins movement to the port. Units traveling outside of CONUS use both sea and aerial ports of embarkation (SPOE/APOE). Heavy equipment and the soldier operators travel by ship to the theater of operations. The remainder of the unit usually travels by air. Splitting the unit presents a challenge for the UMT.

Since the UMT will normally travel by air, RS for soldiers traveling by ship will require special planning and coordination.

Holding Areas

3-72. While the unit awaits transportation to the SPOE/APOE, it is normally confined to a Personnel Holding Area (PHA). Time spent there can seem long and monotonous. If the holding area is on the installation, this can be a frustrating time for soldiers because families are near but out of reach. This can be an important opportunity to provide religious support.

Deployment Phase III: Strategic Lift

3-73. Strategic lift begins with departure from the POE and ends with the arrival in theater. Transportation Component Commands are responsible for the strategic transportation of forces and their support. Religious support during this phase is oriented toward soldiers traveling with UMTs to the theater of operations.

Deployment Phase IV: Theater Base Reception

3-74. Force reception begins when the unit arrives at the port of debarkation (POD) in theater and ends when the unit departs the POD to begin the onward movement phase. One of the primary requirements during this phase is coordinating the onward movement of the deploying forces to their destination. The theater chaplain is responsible for religious support in the ports of debarkation.

3-75. Except in the case of opposed entry, CSS units may arrive early in the deployment to process combat units through the POD and to establish logistical support systems. Religious support resources in the port-based CSS units are limited. To prepare for operations, units are moved out of the PODs as rapidly as possible, through marshaling or staging areas, and into assembly areas (AA) or logistics bases.

3-76. In assembly areas soldiers prepare for onward movement. The unit may task organize the force, develop and issue tactical plans and orders, coordinate with other units, conduct reconnaissance, conduct training, and rehearsals. The immediacy of the pending mission frequently focuses soldiers' concerns upon matters of faith or religion. When units are in Assembly Areas (AA), UMTs are generally able to provide religious support to soldiers with little risk and with great receptivity.

Deployment Phase V: Theater Onward Movement

3-77. Onward movement begins with the personnel and equipment linkup, the reconfiguration of forces, sustainment, and receipt of pre-positioned war reserve stocks at designated areas. This phase concludes with the unit's arrival at the gaining command's Intermediate Staging Base (ISB) and tactical assembly areas.

Intermediate Staging Base

3-78. Units traveling through the ISB en-route to their combat mission present some religious support challenges. ISBs vary in size depending on the forces involved in the mission, and at times can be large in terms of population and geography. Units remain in the ISB for a minimum time, thus creating a highly transient community.

Tactical Assembly Areas

3-79. Tactical assembly areas (TAAs) are occupied by brigades, battalions, and companies that are forward of the rear area. Enemy contact is likely and the commitment of the unit directly into combat is possible and often anticipated.

3-80. Actions in the TAA focus on preparing the unit for future operations. Common TAA activities include task organization, maintenance, personnel replacement, re-supply, reorganization, rest, and planning for future operations. UMTs monitor changes in the task organization and adjust their plans accordingly. TAAs offer good opportunities for religious services.

Continental United States Replacement Centers

3-81. Replacement centers within the Continental United States (CONUS) receive and certify individuals for deployment to a combat theater. The CONUS Replacement Centers (CRCs) process non-unit related AC soldiers (to include RC soldiers accessioned onto active duty), Department of the Army civilians, contract civilians, Red Cross workers and other civilians.

BATTLEFIELD TRANSITION PLANNING AND EXECUTION

3-82. As with planning for different types of operations, planning for mission transitions must extend throughout the planning process and into operations and redeployment. Transferring control of an operation to another organization requires detail planning and execution. Transition depends upon the situation and will have its own characteristics and requirements. The UMTs should continuously communicate and coordinate among themselves, ensuring continual RS. After a hostile conflict or engagement, regeneration of force capabilities will be a primary consideration for the UMT in the transition plan. Following a stability or support operations mission, the soldier's physical, emotional, family, and spiritual recovery may be the UMT's transition focus. Mission analysis should provide the initial information to commence transition planning. The RSP supports the command's OPLAN with thorough integration. Transition planning and execution considerations include—

- Logistical supplies.
- Synchronized RSP transition with follow-on units.
- Denominational support.
- Transition religious support continuity file.
- Religious and Cultural information of the AO.

3-83. A smooth transition between forces and UMTs is essential for continuous religious support of all the military units. Cooperation among UMTs will ensure that all military and approved personnel receive quality religious support.

RELIGIOUS SUPPORT DURING REDEPLOYMENT AND DEMOBILIZATION

3-84. With the cessation of hostilities or the declaration of a truce, the deployed force in a theater of operations will transition through the last stages of Force Projection. This transition from the post-conflict stage through the redeployment and demobilization stages may begin even if residual combat operations are still underway.

3-85. While some forces are preparing for redeployment, other forces remain in support of Peace Building Operations. While soldiers restore order and reestablish the host nation infrastructure, UMTs provide religious support during the post-conflict stage.

3-86. As operations cease and forces are no longer required for Peace Building, units begin the redeployment stage. Forces and materiel not required for subsequent operations re-deploy to CONUS or home-theater to prepare for future missions. Redeploying forces may deploy to another theater or to an area other than home station on short notice. Redeployment, like deployment, occurs in several phases:

- Reconstitution for strategic movement.
- Movement to redeployment assembly areas.
- Movement to the port of embarkation.
- Strategic lift.
- Reception at a port of debarkation.
- Onward movement from port of debarkation.

3-87. During demobilization the process of transferring RC units, individuals, and materiel from an active posture to a pre-mobilization or other approved posture begins. Demobilization coincides with the first three phases of redeployment, the phases before the units depart the theater of operations. During this stage, RC units return to pre-mobilization status.

3-88. Demobilization requires careful planning which takes into account public relations and what the soldiers are thinking and saying to each other.

3-89. The phases of demobilization are—

- Planning.
- Area of operations demobilization actions.
- Port of embarkation to demobilization station operations.
- Demobilization station operations.
- Home station actions.

Unit Religious Support During Redeployment

3-90. Religious support during redeployment is similar to religious support during deployment. Some soldiers and contractors will be preparing equipment for strategic lift or waiting in holding areas for transportation, while others will be engaged in force projection operations.

> Worship, pastoral care, and religious education help soldiers begin the transition back to normal life.

3-91. UMTs providing religious support in assembly and PHAs during redeployment face challenges similar to those encountered during deployment. Worship, pastoral care, and religious education will help soldiers begin the transition back to normal life. Reunion sessions, an important part of the UMT's work, require early planning to be effective.

Reunion

3-92. Reunion denotes the reuniting of family members after separation as a result of deployment. Army research shows that family problems intensify when spouses return from deployments. Characteristics of the readjustment period, which lasts for about two months, include the emergence of physical symptoms, children's behavior problems, and marital tensions. For many couples, the first weeks of reunion are the most challenging part of the whole deployment experience. Tensions become most acute during the two weeks before and after the soldiers' return.

3-93. Religious support during reunion assists soldiers and family members in dealing with the stress that may occur while attempting to integrate experiences while separated, as well as, personal expectations of family life after deployment. UMTs help to restore and preserve the relationships that couples and families knew before deployment. During the reunion, UMTs join with the Installation Religious Team, Army Community Services, STARCs, United States Army Reserve Command (USARC) Major Subordinate Commands (MSC), family program coordinators, and mental health personnel to provide family assistance.

> By assisting with the stress of reunion, the UMT helps to restore and preserve the relationships that couples and families knew before deployment.

3-94. As married soldiers encounter reunion issues with spouses, single soldiers will experience reunion struggles with family and friends. Soldiers deployed as individuals or in small detachments also experience reunion problems upon return to their units. Reunion briefings and workshops help to minimize the problems for soldiers, families, and units. These events equip soldiers with coping skills to face the challenge of reintegration to families, peer groups, and units.

Reunion Preparation In Theater Before Redeployment

3-95. Before departing a theater of operations, UMTs coordinate with the S1/G1 for reunion briefings and workshops. The workshops teach skills needed to reconnect with younger children or to bond with newborns. During the pre-reunion period soldiers and families wonder, "What will it be like to be together again?" and "How has this experience changed us and our relationship?" While deployed UMTs conduct workshops in theater, non-deployed UMTs conduct similar workshops for waiting families.

> UMTs coordinate with the S1/G1 for briefings and workshops for soldiers in order to prepare them for reunion.

Reunion Preparation at Home Station after Redeployment

3-96. UMTs plan additional workshops at home station. Some workshops will be for soldiers and spouses, others for soldiers or spouses alone. UMTs may coordinate workshops with the Installation Ministry Team (IMT), ACS, Family Readiness Groups (FRG), and Chaplain Family Life Centers. The UMT must be prepared to sustain the effort for several months, depending on the length of the separation and the severity of the mission. From time to time, UMTs conduct assessments of the effectiveness of the workshops and the need to continue them.

3-97. Members of UMTs are cautioned not to ignore their own family reunion process. Being occupied with helping others is not a reason for neglecting one's own family. Installation and Family Life Chaplains conduct workshops specifically for redeploying UMTs.

UMTs are cautioned not to ignore their own family reunion process.

Recovery/Refitting

3-98. The unit's need to reestablish its readiness level sometimes competes with the soldier's family for the soldier's time, energy, and attention. During the recovery and refitting phase, soldiers need to clean, repair, and secure equipment; identify shortages; requisition supplies; and reestablish stock levels. Property accountability is reestablished with the installation Property Book Officer (PBO). UMTs are not exempt from these concerns, even though they are busy helping soldiers and families to reintegrate.

UMTs must balance reunion activities with refitting.

RELIGIOUS SUPPORT DURING DEMOBILIZATION

Planning Actions

3-99. Demobilization planning begins with mobilization and ends with the decision to release units and individuals from active duty. This planning includes—

- Relocating units and individuals.
- Incorporating annual training (AT) for non-mobilized RC units to support demobilization.
- Mobilizing additional RC units.
- Using individual volunteers (by means of temporary tours of active duty) to support critical missions in the demobilization process.
- Basic maintenance of equipment.
- Awards and decorations.
- Evaluation reports.
- After Action Reviews/Lessons Learned.
- Chaplaincy Detachment (CD) and GSU integration for support.

In Theater Activities

3-100. Before a unit departs the theater, it initiates demobilization administrative and logistical actions to include—

- Processing awards, decorations, and performance reports.
- Initiating line-of-duty investigations.
- Updating supply records and equipment accountability.

3-101. Medical and dental actions are started in theater when resources are available. During this phase soldiers may be frustrated as they become anxious to return home. As always, UMTs adapt religious support based on an assessment of soldiers' needs.

Home Station Activities

3-102. The reintegration of soldiers who deployed with those who did not is an important consideration for UMTs. During mobilization, many soldiers may deploy at different times, or with other units as a result of cross leveling. During demobilization, they may also return to home station at different times. Feelings of resentment may build in the unit, and in families, because some soldiers have returned and others have not. Reentry into the civilian work force is of another concern of the soldiers of the RC.

INSTALLATION RELIGIOUS SUPPORT

REINTEGRATION/REUNION

3-103. The Installation Religious Team plans for—

- The reintegration of returning soldiers with their family members and with non-deployed soldiers in their units.
- Support to casualty survivors.
- Non-deploying UMTs to continue their coverage of units and family members normally served by the returning UMTs, so that they will be able to reintegrate without the pressure of immediate duty.

3-104. Chaplain Family Life Center will—

- Assist in preparing families for reunion with soldiers.
- Assist UMTs with training and technical guidance on reunions.

(See Chapter 7 for further discussion of the Installation Chaplain and NCOIC responsibilities.)

GENERAL RESPONSIBILITIES OF THE UNIT MINISTRY TEAM DURING RELIGIOUS SUPPORT OPERATIONS

Staff Supervision

3-105. Chaplains at brigade and higher levels provide staff supervision of religious support by visiting subordinate units, monitoring the tactical situation, and revising estimates and plans based on changes in METT-TC.

3-106. A staff officer does not have the authority to order a subordinate unit to execute directives or orders. Staff officers make recommendations to subordinate commanders. These recommendations may be accepted or rejected. A staff chaplain may contact a subordinate UMT to provide instructions, advice, and recommendations, offer assistance, or exchange information. The commander will establish the nature and degree of the chaplain's actual authority or control in implementing the RSP.

Advising

3-107. Chaplains advise the commander on religion, morals, ethics, and morale as affected by religion. Advice extends to concerns such as the quality of life for soldiers and families, impact of indigenous religions on operations, and the allocation of resources. Commanders rely on the advice of the UMT to help them discern the unit climate. At brigade and higher levels, the UMT advises the command on the assignment of chaplains and chaplain assistants. The chaplain's direct access to the commander allows face-to-face communication and is the most effective means for providing advice.

Administration

3-108. Administrative actions continue in the tactical environment. They comply with unit TSOPs and appropriate regulations and doctrine.

Correspondence

3-109. Battlefield conditions complicate the preparation of correspondence. The necessity for correspondence, however, will not decrease. The UMT must prepare required reports and correspondence despite equipment limitations or the tactical situation. Prior to deployment, UMTs must consider how to adapt their procedures to field conditions.

Records Management

3-110. Army regulations and the unit TSOP determine the procedures for managing and maintaining records. The UMT may need to maintain the following records:

- Baptisms and other sacramental acts.
- Services of worship and attendance.
- Ministration to casualties.
- Memorial services or ceremonies.
- Staff journal (See DA Form 1594).
- After Action Reviews.

Logistics

3-111. Logistic support is essential to religious support in combat. Because support requirements are more difficult to meet in combat, UMTs must become familiar with the military supply system and use it as their first source of re-supply rather than making local purchases. UMTs should know what items are available within the system, which items are authorized for the UMT, and what procedures to use to requisition supplies.

3-112. The Modified Table of Organization and Equipment (MTOE) is the authorization document for ordering equipment. The chaplain is responsible for all team equipment; while the chaplain assistant is accountable for its proper use, security, and operational readiness. Request for changes to the MTOE should be coordinated and staffed through the chain of command using DA Form 2028 (Recommended Changes To Publications and Blank Forms).

Supplies

3-113. The chaplain assistant monitors supply levels, conducts inventories, and requisitions supplies through unit supply channels digitally using Department of Defense on-line requisition sites, or in coordination with the Defense Supply Center Philadelphia (DSCP) Item Manager. This procedure will ensure timely replacement of standard expendable items. The team reports its supply requirements to the supply section by national stock number (NSN), nomenclature, amount, and projected usage rates.

3-114. UMTs should deploy with at least 60 days of ecclesiastical supplies. Normally the unit establishes stockage levels, expressed as "Days of Supply" (DOS). The UMT may need to adjust the amount of supplies carried in accordance with the projected rate of resupply for each mission. A team deploying to an immature theater with limited logistical support may not have resupply available for 90-120 days.

3-115. UMTs at division and higher echelons coordinate directly with the Defense Logistics Agency (DLA) Liaison Officer and the 9925 Item Manager at DSCP to ensure a steady flow of Chaplain Resupply Kits and other religious support supplies to subordinate teams.

3-116. Planning must incorporate religious education/sustainment supplies for long-term continual operations. These supply items are also managed by the 9925 Item Manager. The supplies will support all faith groups' spiritual sustainment needs.

Maintenance

3-117. Maintenance is critical in a combat environment. The team is responsible for the maintenance of its assigned vehicle and other equipment. An effective maintenance program is essential to ensure the UMT's ability to perform religious support. Operator maintenance focuses on preventive maintenance checks and services (PMCS). Both the chaplain and the chaplain assistant must be able to operate and maintain the vehicle. These skills must be acquired and practiced before deployment.

Transportation

3-118. Section 3547 of 10 USC mandates transportation assets be afforded to the Chaplain for the conducting of religious services. The vehicle must be able to transport the team to all unit locations within the AO. The team must take precautions to avoid compromising the unit's location or fighting positions. Both members of the UMT must be trained and licensed to operate the vehicle.

3-119. METT-TC will determine who drives the vehicle. The practice of the chaplain driving in hostile areas ensures that the assistant, who is a combatant, can provide security. When not providing security for the team, the chaplain assistant operates the vehicle.

3-120. The UMT should take full advantage of unit convoys. It must be thoroughly familiar with convoy procedures, unit policy regarding movements, and vehicle densities.

Communication

3-121. As the UMT travels on the battlefield, it must maintain communications with its operational base to keep the base informed of its status, location and travel plans. At the same time, the team receives and transmits situation reports (SITREP). The commander must provide the team with dedicated communications equipment that is compatible with and keeps pace with the commands secure communication equipment. The team may use a variety of other means available within the unit. The team must be flexible and creative to maintain communications.

3-122. Religious support is synchronized and implemented throughout the area of operations, theater of operation, and communications zone (COMMZ) by being fully integrated into the digitized communication and automation systems. The ability to reach from the area of operations to the home station support node (HSSN) is essential for relaying time sensitive information and religious support requests. A mobile communication link between each echelon enables the UMT to allow time-sensitive data to be rapidly communicated to and from each echelon. This is essential for the UMT's mission of providing soldiers with the personal delivery of religious support. It ensures that the UMT has the ability to move to where time-critical religious support is essential. The communication linkage between UMTs provides immediate access to processed critical information needed to ensure the commander's religious support plan is executed successfully.

Assessment

3-123. UMTs continuously identify, assess, and disseminate lessons learned throughout the force and branch. Assessment of lessons learned from operations is the key element for future training objectives. Observation teams deploy with Army forces on every major operation. Religious support is an important element of the operation and lessons learned must be gathered on and integrated into the overall assessment of the operation.

3-124. Lessons learned are also critical during a transition phase between one force and another. UMTs arriving into the theater of operations can better understand the battlefield and METT-TC implications through lessons gathered by departing UMTs.

RELIGION, MORALS, ETHICS, AND MORALE

3-125. Chaplains serve as the "conscience of the command." Chaplains advise the commander on the moral and ethical nature of command policies, programs and actions, as well as, on the impact of command policies on soldiers. They advise the commander on—

- Religious/ethical issues as they bear on mission accomplishment and morale.
- Inter-faith relations among religious groups in the command.
- The role of religion within indigenous population groups.
- Accommodation of special religious needs/requirements.
- Quality of life issues.
- Unit command climate.

INDIGENOUS RELIGIONS

3-126. The chaplain advises the commander and staff on the beliefs, practices, and customs of religious groups in the AO, and the impact that local religions may have on the conduct of the mission. This advice is based on information developed at division and higher echelons and is coordinated with the G5 and Civil Military Operations (CMO) officer. (See Appendices A and F for more information.) Common concerns include—

- Religious organizations and doctrines.
- Religious practices and customs.
- Places of worship, shrines, and other holy places.

PUBLICATIONS

3-127. Prior to deploying, UMTs review the list of publications in the reference section of this manual to determine what to carry. As a minimum, the team carries this manual (FM 1-05), JP 1-05, AR 165-1, and pertinent "go to war" publications of the unit. The chaplain assistant requisitions missing publications and maintains the team's publications.

JOURNALS AND WORKBOOKS

3-128. The UMT should maintain a staff journal covering each 24-hour period (See DA Form 1594). The staff journal is a chronological record of events and serves as a permanent record for operational reviews, training issues, and historical research. The team's journal entries should provide enough detail to describe its activities adequately and to fix the time and place of significant religious support events; however, care must be taken to avoid compromising privileged information by recording it in the staff journal. Journals are useful in reconstructing events, maintaining religious support histories, writing AARs and lessons learned for the operation. Typical entries include—

- Summaries of plans and orders.
- Religious support activities.
- Reports sent and received.
- Reports of contact with other UMTs, missionaries, faith-based NGO/IOs, and humanitarian agencies.

3-129. A UMT may choose to keep a staff section workbook, also known as a "battle book." It is an informal, indexed collection of references and information taken from written and oral orders, messages, journal entries, and meetings. The battle book may include extracts from doctrinal manuals, forms, and other information. It has no specific format and is not a permanent record. (See FM 101-5.) RC UMTs should build battle books in concert with the AC unit they train with in peacetime and/or deploy with for operations.

SUMMARY

3-130. Religious support operations planning, preparation, coordination, and execution must be comprehensive. It must envelop all faith groups and personnel, be synchronized among all commands and UMTs, and be flexible enough to meet the fluid environment. The UMT will be faced with a great deal of information it must manage, implement, and disseminate. Constant and aggressive information gathering and analysis will keep the UMT informed of the religious support requirements for the commander. The chaplain, as a personal staff officer, and the chaplain assistant are a critical assets for ensuring the commander's religious support plan is implemented in a comprehensive and timely manner. Through proper staff procedures, mission rehearsals, and staff coordination, the RSP will meet the mission's requirements.

RESOURCES

3-131. The following is a list of main resources utilized for writing and updating this doctrine.

- AR 165-1.
- JP 1-05.
- JP 5-00.2.
- JP 3-0.

- JP 3-05.
- FM 16-1.
- FM 3-0 (100-5).
- FM 25-100.
- FM 25-101.
- FM 100-17-3.
- FM 101-5.
- FM 55-30.
- National Military Strategy
- Joint Vision 2010
- Army Vision 2010
- Concept for Future Joint Operations, May 1997
- U.S. Army Chaplaincy Strategic Plan FY 2000-FY 2005
- Chaplain Support Force XXI White Paper
- Quadrennial Defense Review Report, September 2001.

Chapter 4

BATTALION, SQUADRON, BRIGADE, REGIMENT AND GROUP RELIGIOUS SUPPORT OPERATIONS

The levels described within this chapter are the smallest echelons to which UMTs are assigned.

CONTENTS

TACTICAL ECHELON DESCRIPTION

BATTALION AND SQUADRONS

4-1. Maneuver battalions can be reinforced with other combat and combat support companies to form task forces for special missions. Task organization increases the capability of maneuver battalions to ensure operational success. A Brigade commander through the staff estimation process may task-organize tank and mechanized infantry battalions by cross-attaching companies between these units. Field artillery (FA) batteries can be reinforced with batteries of any kind to form artillery task forces. Light units and engineer battalions can also be reinforced using this same task organization process.

4-2. Combat support and CSS battalions vary widely in type and organization. They may be separate division or non-divisional battalions but normally perform functional services for a larger unit within that unit's area of operations. Air defense artillery and signal battalions assigned to, or supporting, divisions routinely operate throughout the division area of operations. The Corps Support Battalion (CSB) is part of a Corps Support Command (COSCOM) and a Corps Support Group (CSG) and provides support to units that could belong to a division, or a Corps.

4-3. Battalions, or their equivalent, conduct operations as part of a brigade's operation. Occasionally they are directly under a division or regiment's control. Their capability is increased through task organization. The commander understands the intent of the brigade commander and division commander so that he can properly employ his force.

BRIGADES, REGIMENTS, AND GROUPS

4-4. Brigade size units control two or more battalions or squadrons. Separate brigades of infantry, armor, field artillery, air defense artillery, engineer, and aviation, and armored cavalry regiments can be used to reinforce corps or division and can be shifted from unit to unit to tailor forces for combat.

4-5. Division commanders adjust the organization of their brigades and change their task organization as required by the factors of METT-TC. They normally receive support from FA battalions, engineer battalions (heavy divisions), forward support battalions, and smaller combat, combat support, and CSS units. Their chief tactical responsibility is synchronizing the plans and actions of their subordinate units to accomplish tasks for the division or corps.

4-6. Separate maneuver brigades and armored cavalry regiments have a fixed organization with organic cavalry, engineer, air defense, field artillery, military intelligence, chemical, and CSS units. Separate brigades are capable of operating as independent units.

4-7. Other combat, CS, and CSS brigades and groups are organized to control capabilities for divisions, corps, and larger units. They may also be the building blocks of large unit support structures such as corps and theater army support commands and of combat support commands such as engineer commands. Divisions receive support from an organic brigade-sized, division artillery (DIVARTY), an aviation brigade, and a Division Support Command (DISCOM) comprised of CSS battalions and companies. Heavy divisions have an organic engineer brigade.

Part of a Division

4-8. Brigades or their equivalent usually operate as part of a division. The division is a large, fixed Army organization that trains and fights as a tactical team. Normally, the division commander controls two- to five-ground maneuver brigades. The brigade can be employed in autonomous or semiautonomous operations when properly organized for combat. All brigades must be able to deploy, conduct offensive operations, conduct defensive operations, and conduct retrograde op-

Planning Checklist

Staff: communication, annex, integration, briefings
Specialized:
• Nurture the living:
  ◆ Spiritual Fitness
  ◆ Worship
  ◆ Counseling
  ◆ Visitation
• Care for the Dying:
  ◆ Spiritual triage
  ◆ Leadership
  ◆ Battle Stress
• Honor the Dead:
  ◆ Memorial Services
  ◆ Memorial Ceremonies
  ◆ Rites
  ◆ Sacraments
  ◆ Prayers

erations. Airborne, air assault, and ranger brigades/regiments are capable of conducting forced entry operations. All brigades may deploy to conduct military operations other than war. The brigade UMT is part of the division religious support staff and program.

Part of a Joint Task Force

4-9. Brigades or similar may deploy as part of a JTF with or without its traditional divisional headquarters. In these types of operations, the brigade may work directly for the JTF commander. The demand for experienced liaison will be high for a brigade in a JTF. Liaison may be required with joint, multinational, interagency, and or non-governmental agencies. The brigade's requirement for liaison will exceed its normal personnel and equipment structure.

4-10. Planning Religious Support for this type of operation needs to take into consideration the diversity and complexity of the chain of command, logistics, force structure, and area of operation.

FUNDAMENTALS OF RELIGIOUS SUPPORT OPERATIONS

GOAL OF UNIT MINISTRY TEAMS

4-11. The goal of all Unit Ministry Teams is the personal delivery of comprehensive, timely, and effective religious support to all soldiers.

4-12. Comprehensive, because each soldier has his or her own definition of what constitutes religious support. While not every spiritual need of every soldier can be met in the battlespace, the UMT seeks to meet as many needs as possible. Timely, because time is a limited commodity, and the situation changes in the battlespace. The UMT must be aware of the changing situation and prioritize ministry accordingly. Effective, because religious support must be more than minimal presence and a hurried service by the UMT. Quality effective religious support can provide hope and courage to soldiers who are facing the stress of battle.

ROLE OF THE BATTALION/SQUADRON UNIT MINISTRY TEAM AND SECTION

BATTALION/SQUADRON TACTICAL UMT	
CHAPLAIN (O-3)	**CHAPLAIN ASSISTANT (E-3/4/5)**
Role Commissioned Officer/Staff Religious Professional Title 10 Protected	**Role** Noncommissioned Officer Specialized Assistant Religious Support Specialist Team Security
Mission Advises commander on religious support and quality of life issues to support the unit's mission.	**Mission** Advises Chaplain and NCO Chain. Provides specialized assistance in areas of religious support to support the chaplain.

BATTALION/SQUADRON TACTICAL UMT	
CHAPLAIN (O-3)	**CHAPLAIN ASSISTANT (E-3/4/5)**
Performs/provides/coordinates the religious support mission for the battalion/squadron. Team leader.	Perform battle staff functions. Team member/ NCO.
Parallel Tasks Performs religious support. Provides spiritual, stress, fatigue, suicide preventive counseling to individuals and family members. Plans unit CMRP. Initiates religious support plan. Manages section funds and field offerings. Supervises Chaplain Assistant/Team NCO.	**Parallel Tasks** Prepare and synchronizes religious support and provides emergency religious ministrations in the battlespace. Screens stress, fatigue, suicide individuals for counseling or provide referrals if needed. Implements the unit CMRP. Prepares the Religious Support Annex. Safeguard and accounts for field offerings. Maintains unit equipment.

RESPONSIBILITY AND ACCOUNTABILITY

Staff Responsibilities

4-13. The chaplain serves the commander as a personal staff officer to plan, synchronize, and coordinate religious support, within the commander's area of responsibility. The battalion chaplain is the section leader. The chaplain assistant serves as the section NCOIC.

4-14. Chaplains serve as the "conscience of the command." Chaplains advise the commander on the moral and ethical nature of command policies, programs, actions, and the impact of command policies on soldiers. They advise the commander on—

- Religious and ethical issues as they effect mission accomplishment.
- Indigenous religions and their impact on the mission.
- Accommodation of special religious needs and practices within the command.
- Matters of unit morale as effected by religion.

Relationship to Brigade Unit Ministry Team

4-15. The battalion/squadron UMT works closely with the brigade (BDE) UMT ensuring a comprehensive and coordinated religious support plan for the BDE commander. The BDE UMT technically supervises the subordinate UMT and supports their RSP. During combat operations the battalion (BN) UMTs maintain communications with the BDE UMT sharing information about the religious support mission. In garrison operations, the BN UMT trains with the BDE UMT, coordinates unit and family religious support, and prepares for deployments.

Relationship to the Division Unit Ministry Team

4-16. The battalion/squadron UMT looks to the division UMT as their mentors and advisors during garrison operations. During combat operations, the

BN UMT will be prepared to communicate religious support issues directly if they assume the BDE UMT mission or if they work directly for a division/task force (TF) UMT. The principle of understanding and implementing the guidance of the commander two levels up also applies to the work of the Battalion UMT. The BCT UMT plans its RSP in harmony with the guidance the Division UMT. The Battalion UMT must understand and implement the Division RSP as it plans and executes its RSP.

Supervision Responsibilities

4-17. The chaplain exercises staff supervision and technical control for religious support throughout the battalion AO. The Chaplain is responsible for the development, implementation, and coordination of religious support assets, operations and activities. The chaplain assistant NCOIC is an active member of the NCO support channel in the unit on religious support issues and soldier welfare. The Chaplain assesses the soldiers and other authorized personnel's interpersonal/spiritual well being, which can affect the unit's fighting spirit. The UMT advises the commander on humanitarian aspects of operations, and the impact of command policies on indigenous religions. The Chaplain provides and coordinates privileged and sensitive spiritual counseling and pastoral care to the unit's command, soldiers, authorized civilians, and families. The UMT locates where they can best coordinate, communicate and facilitate religious support across the battlespace.

Staff Supervision During the Operation

4-18. The battalion Chaplain supervises the Chaplain Assistant. The UMT provides direct support and area support to units located within the battalion AO. The UMT coordinates religious support with each unit. The UMT monitors the tactical situation, and plans based on changes in METT-TC. If necessary, they recommend changes to the religious support annex and ensure supported units receive information concerning changes via` FRAGOs. The UMT reports required religious support data to the brigade UMT. The battalion UMT receives operational missions via the Brigade Operational Order/FRAGOs. The brigade UMT is the technical supervisor for the battalion UMT who does prior coordination and establishes contingency plans for the brigade religious support plan.

4-19. Before the battle, the battalion UMT submits a religious support annex that gives guidance to all unit commanders of the religious support provided. The chaplain assistant gathers information that will allow the UMT to battle track the operation, current unit positions, overlays, special operating instruction (SOI) information and successfully navigate the battlespace.

4-20. The UMT before, during, and after a phase of operation will coordinate with the Battalion Task Force for treatment and forward surgery teams to provide comprehensive religious support to soldiers being treated and to the team itself. The UMT coordinates their efforts in conjunction with the movement of the casualties. The UMT maintains contact via radio, casualty feeder cards, etc. with the brigade UMT in the form of SITREPs on their location, health, statistics, and ability to provide religious support such as mass casualty (MASCAL).

4-21. The battalion UMT considers the spiritual condition of the soldiers in follow-on planning. The severity of the previous operation and the number of casualties suffered are key factors in planning, preparing, and executing cycle that is a nonstop process.

Training

4-22. The battalion UMT has the responsibility to be proficiently trained in all areas necessary in order to adequately execute the religious support mission. It is the responsibility of each member of the UMT to be a tactically and professionally proficient team member. They must train in soldier tasks and collective tasks as a team. The battalion UMT trains to the unit's METL utilizing the derived religious support METL. The religious support METL serves as the focal point on which the UMT plans, executes, and assesses training. Training must encompass the following principles to effectively train as a team:

- Train as you provide religious support in the battlespace.
- Use all appropriate doctrine and tactics, techniques, and procedures (TTPs).
- Use performance-oriented training.
- Use mission-oriented training.
- Make training a challenge.
- Train to sustain proficiency.
- Train using multi-echelon techniques.
- Take control and maximize training opportunities.

ROLE OF THE BRIGADE UNIT MINISTRY TEAM

4-23. The religious support section is made up of a minimum of a Unit Ministry Team (UMT) composed of a chaplain (56A) and one enlisted chaplain assistant (56M). The UMT facilitates and coordinates religious support across the battlespace.

4-24. The brigade chaplain is a religious leader, competent staff officer and team leader. The brigade chaplain assistant NCOIC is a leader and advisor, mentor and information manager for the UMT. The UMT must have a keen awareness of the world situation as it bears on potential military operations, must know the role of military religious leaders and must have a broad understanding of the impact of indigenous religions on military operations.

4-25. UMTs must use technology to enhance the personal delivery of religious support. They must be trained in joint and multinational operations and skilled in synchronizing all aspects of religious support during military operations. They will operate in an environment of increasing volatility, complexity and ambiguity, thus the brigade UMT must be innovative and disciplined.

4-26. The brigade chaplain assistant needs to be a graduate of the Battle Staff NCO Course (with the 2S identifier). This will enable the brigade chaplain assistant NCOIC to take part in the MDMP in the absence of the chaplain, to track the battle analytically, and to better utilize the digital enablers that the brigade will share with other staff sections.

4-27. Brigade UMTs require the same level of technology as the command. UMT automation, communication and transportation must keep pace with the command's requirements in order to effectively coordinate religious support, communicate critical information and remain physically flexible to meet requirements across the brigade battlespace. The UMT is an integral part of the brigade.

THE BRIGADE/REGIMENTS/GROUPS TACTICAL UMT	
CHAPLAIN (O-4/5)	**CHAPLAIN ASSISTANT NCO (E-6/7)**
Role Commissioned Officer/Staff Religious Professional Title 10 Protected	**Role** Noncommissioned Officer Specialized Assistant Religious Support Specialist Team Security
Mission Advises commander on religious support and quality of life issues to support the unit's mission. Performs/provides/coordinates the religious support mission for the brigade. Brigade Team Leader.	**Mission** Advises Chaplain NCO Support Channel Provides specialized assistance in areas of religious support to support the chaplain. Performs Battle Staff functions. Team NCOIC.
Parallel Tasks Provides religious support supervision of Bn/Sqd Chaplains. Initiates and develops the religious support plan. Train Chaplains on MOS task. Plan and reviews the BDE CMRP.	**Parallel Tasks** Provide religious support supervision of Bn/Sqd Chaplain Assistant NCOs and Chaplain Assistants. Synchronizes the religious support plan. Trains Chaplain Assistants on MOS task. Prepare and implement the BDE CMRP.

RESPONSIBILITY AND ACCOUNTABILITY

Staff Responsibilities

4-28. The Brigade UMT works directly for the commander. The chaplain serves the commander as a personal staff officer to plan, synchronize, and coordinate religious support, within the commander's area of responsibility. The brigade chaplain is the section leader. The chaplain assistant serves as the section NCOIC. The brigade UMT serves as the technical supervisor and mentor to subordinate UMTs. As the commander's personal staff officer for religious support, chaplains at brigade-level and higher are responsible for assessing the technical proficiency of UMTs in subordinate units. They conduct assessments with the unit during field training exercises (FTXs), Battle Command Training Program (BCTP) events, external Army Training and Evaluation Program (ARTEP) events, Combat Training Center (CTC) rotations, unit training, and command inspections.

4-29. Chaplains serve as the "conscience of the command." Chaplains advise the commander on the moral and ethical nature of command policies, programs and actions, and the impact of command policies on soldiers. They advise the commander on the following, in addition to, the previous battalion/squadron topics:

- UMT personnel issues.
- Division RSP.
- Training issues.

Relationship to the Division Unit Ministry Team

4-30. The Division or TF UMT is the immediate technical supervisor for BDE UMTs assigned or attached operationally to a division or task force. If not assigned or attached, there may be a need operationally for coordination and close communication to ensure comprehensive religious support.

4-31. The BDE UMT or equivalent fully supports the division and/or TF RSP by supervising and providing guidance to BN UMTs RSPs. They work closely supporting the operational information religious support management process for the entire division or task force.

Relationship to CORPS Unit Ministry Team

4-32. The principle of understanding and implementing the guidance of the commander two levels up also applies to the work of the Brigade UMT. It is essential for the BCT UMT to understand and implement the Corps RSP as it plans and executes its RSP.

Staff Supervision During the Operation

4-33. Brigade UMTs provide staff supervision of religious support by visiting subordinate units, monitoring the tactical situation, and planning based on changes in METT-TC. If required, they recommend changes to the religious support annex and ensure subordinate units receive and execute the changes, which are issued as FRAGOs. The brigade UMT should maintain contact with subordinate UMTs by visits, radio spot reports, or any means possible to insure subordinate UMTs status is 100 percent and religious support data is collected and sent to the division UMT.

4-34. The brigade UMT monitors the tactical situation, the status of each battalion UMT, and plans for future operations. If changes in METT-TC warrant adjustments in religious support, the brigade chaplain recommends changes to the brigade commander or executive officer. Reassigning a chaplain or chaplain assistant, or changing the mission of a battalion UMT, requires a FRAGO. The brigade UMT must be prepared to augment TF UMTs in the event of mass casualties. The brigade religious support annex should have a plan to augment the religious support provided at the base support battalion (BSB), forward support medical company (FSMC).

4-35. When the brigade reaches the objective, casualty evacuation increases. The brigade UMT again assesses the requirements for additional religious support in the AO. If the number of casualties is high, a MASCAL is declared. The brigade UMT will issue a FRAGO to notify supporting UMTs. These UMTs will assist the BSB UMT in providing religious support at the FSMC to casualties and by conducting religious support debriefing.

4-36. If a TF UMT has not submitted a current SITREP (See format model in FM 101-5-2), the brigade UMT NCO contacts the team to determine its status. When feasible, the brigade UMT moves forward to the TF UMTs to provide direct spiritual care. Priority goes to the TF executing the main attack. If the brigade continues the mission, time may be very limited. This phase may start with reconnaissance, or it may go directly into deployment for another assault.

4-37. The brigade UMT considers the spiritual condition of the soldiers in follow-on planning. The severity of the previous operation and the number of casualties suffered are key factors in the planning, preparing and executing cycle that is a nonstop process.

4-38. If the brigade is operating as part of a division mission, the brigade UMT of the follow-and-support brigade establishes liaison with the UMT of the lead brigade to facilitate pastoral care casualties and other religious support. The follow-and–support brigade may provide evacuation and treatment for casualties of the lead brigade, so that the lead brigade can continue its tempo. This mission is of particular concern to the brigade UMT because of its complexity, intensity, and momentum.

4-39. In the mobile defense, battalions engaged in offensive, defensive, and retrograde actions present a challenge for the brigade UMT. The UMT plans religious support for each of the units based on METT-TC. The religious support plan is incorporated into the brigade OPLAN as an annex. The area defense differs from the mobile defense in that all units are engaged in defensive operations.

Supervision of Brigade Combat Team (BCT) Religious Support Operations

4-40. All action starts with a vision and that includes religious support. "Vision" is a personal concept of what the organization must be capable of doing by some future point. It is the target. What does the organization need to accomplish today, tomorrow, or years from now to achieve the desired result? How is the organization developed to achieve that capability? Simply put, vision sees the required effort, leadership, and plan to get the job done.

4-41. The transition from battalion to brigade-level religious support requires a transition in focus as well as location. The brigade-level chaplain's focus moves from being primarily an operator to being a staff supervisor of religious support for the commander. A very important part of being a brigade chaplain/supervisor is the development of the vision mentioned above. That vision must broaden to the whole brigade and see supervision as an aspect of religious support proper.

4-42. One component of the vision is certain imperatives for focusing action -- a vision that includes provisions to provide purpose, direction, motivation, and continuity. To provide some insight as to how this applies to the Brigade Unit Ministry Teams, the following few paragraphs will paraphrase FM 22-103, Leadership and Command at Senior Levels.

4-43. Purpose insures there is a clear and worthy object for every action and sacrifice. It provides the UMTs with a reason to withstand the stress of continuous operations. It depends heavily on the ability of the battalion UMTs to understand all requirements. The brigade chaplain does this by giving a sense of priority and importance to necessary tasks. The brigade chaplain explains how the success or failure of the UMTs will impact upon the unit. Teams with a good sense of purpose will succeed in ways that cannot be forecast before the action. Purpose provides the religious support teams of the BCT with goals. It gives the teams a sense of "why" they are doing what they are doing.

4-44. Direction channels the collective efforts of the people on the teams toward the unit's goals, objectives, and priorities. It puts the Unit Ministry Teams on a clear and unambiguous course and thus creates the conditions that allow them to succeed. Setting goals and standards, ensuring discipline, and training the teams are some of the means the brigade chaplain can use to chart a direction for the teams. Goals and standards provide the azimuth. Good discipline provides the confidence that goals and standards are met in the absence of leaders and immediate guidance. Training ensures results. An additional means of providing direction is the SOP. Direction helps the teams know what to do, while purpose helps the teams understand why they do it.

4-45. Motivation engages the will of Unit Ministry Teams to do their utmost to achieve the purposes, goals, objectives, and priorities of the unit, even under the most difficult circumstances of combat. It is not enough to know why and what to do; we need the moral force of a strong will.

4-46. Continuity anticipates future conditions, situations, and needs; it takes steps in the near term to ensure that the BDE Unit Ministry Team can meet future missions. Synchronization of plans and efforts among all Unit Ministry Teams is critical to quality religious support. The Brigade Chaplain is the focal point for this as the staff representative for religious support for the commander.

Supervisory Relationships

4-47. The complex job of the brigade chaplain is to organize the efforts of UMTs in different subunit commands with different battlefield operating system (BOS) responsibilities and support. Some of the teams are in units assigned directly or as attached to the brigade, i.e., the maneuver battalions and field artillery battalion (unit support). To further expand religious support supervision and synchronization, area support to soldiers, members of other services, authorized civilians, and all who have no organic or available religious support is part of the comprehensive plan. The supervisory role expands to denominational coverage to these soldiers. Still, the mandate to UMTs is to provide religious support to all soldiers in the AO.

4-48. The Brigade Chaplain as a staff officer for the commander uses different external coordination channels to ensure the accomplishment of religious support. Three channels of communication command, staff, and technical require development and rehearsal. Too often coordination and guidance are communicated through the wrong channels; and because the Brigade Chaplain has no command authority, he/she must work effectively through the correct coordination channels.

Supervision of Brigade Religious Support Operations

4-49. Task Force UMTs work for their unit commanders. The BDE UMT cannot give guidance to subordinate UMTs that conflicts with their commander's guidance. Proper coordination and solid common sense on the part of the BDE UMT make the entire BDE religious support plan work.

4-50. Proper coordination with the commands and staffs in the brigade opens doors of quality communication and support. A good working relationship builds respect and cooperation with commanders and staffs throughout the brigade combat team. The brigade chaplain and NCOIC need to cultivate this relationship and respect their position on the commander's staff. Proper staffing procedures make religious support a reality.

OBSERVED PRINCIPLES OF EFFECTIVE SUPERVISION

The primary mission of the Brigade UMT is to synchronize Religious Support.

The BDE Unit Ministry Team assigns its highest priority to managing religious support for the whole BDE. The UMT must see supervising as their part of comprehensive religious support. Providing religious support to the tactical operations center (TOC) is secondary to the primary mission of synchronizing religious support throughout the BDE. When the BDE UMT is effectively overseeing religious support, quality religious support reaches all corners of the BDE.

PRINCIPLE 1. The BDE UMT trains as a team before it arrives in a theater of operation.

A goal of the BDE UMT training is to work as a synchronized team. UMTs do individual religious support well, but the BDE UMT must work hard at coordinating the religious support that happens in the Brigade.

PRINCIPLE 2. Synchronize Religious Support Appendices at all levels.

The BDE UMT ensures that religious support is synchronized on paper and in action. Make certain that the Brigade Religious Support Appendix is accurately reflected in subordinate UMT Religious Support Appendices. A goal for BDE UMT training is to have continuity in religious support appendices throughout the BDE UMT.

PRINCIPLE 3. Plan, rehearse and coordinate casualty replacement plans for members of the BDE UMT.

UMT members in the BDE should know who would replace them in the event they become a casualty.

PRINCIPLE 4. Coordinate all Religious Support guidance with the appropriate commanders.

Just because something is written in an appendix does not mean that it will happen, it must be coordinated. Coordinate religious support prior to the mission with all subunit UMTs. This effort helps to ensure comprehensive religious support synchronization in the AO.

OBSERVED PRINCIPLES OF EFFECTIVE SUPERVISION
PRINCIPLE 5. Daily communication is vital to Religious Support. UMTs have a crucial piece of the overall religious support plan and need to send that information up and down. UMTs must know what sort of information is necessary to report. BDE UMTs must be aggressive in maintaining disciplined and systematic information flow. Daily communication between all UMTs in the BDE is imperative for up-to-date information. Here are some suggestions for making that work: • Net calls. A net call establishes a specified frequency and time when all UMTs are on the radio at the same time. In this way information can be put out quickly and every UMT becomes knowledgeable of the religious support situation. • Maintain regular hours for monitoring the radio and mobile subscriber equipment (MSE) so subordinate UMTs know when they can directly contact the BDE UMT. • Send daily guidance and updates to subordinate UMTs. • Spot reports to pass on time sensitive information. • Use the BDE RETRANS frequency (coordinate with the signal officer. • Send messages by couriers such as commander's drivers, ambulance drivers, and logistics package (LOGPAC) drivers. • Authorize lateral communication between UMTs. Often one UMT cannot make radio-communication with the BDE UMT, but can pass a message through a neighbor UMT. • Schedule regular meetings with subordinates in the field.
PRINCIPLE 6. Daily updates from the BCT UMT to subordinate UMTs are critical. When the BDE UMT communicates to its subordinate UMTs it builds teamwork. This enables individual UMTs to understand what is happening, and what is expected of them. They feel like part of the team. Always remember that effective communication is a two way street. UMTs must communicate the following items of information: • A daily status report of all UMTs in the BDE. • UMT NCOs update. • Changes in religious support for attached/detached units, and units operating in the brigade AO. • Upcoming religious events in the command, and indigenous religious holidays. • The Brigade Commander's daily guidance and other important issues from the command. • Questions that need answering reference indigenous religions and their potential impact on operations.
PRINCIPLE 7. Daily SITREPs from subunit UMTs to the BDE UMT are critical. The BDE UMT must know what the subunit UMTs know to effectively manage the religious support mission. Emphasis upon effective communication will build teamwork and enhance the religious support mission. Examples of information that must be communicated follows: • The status of the UMT. • The status of ecclesiastical supply. • Daily activity. • Changes in unit RPP. • CA issues. • Information on indigenous religious leaders and sacred sites. • Issues of religion, morals, ethics, and morale.

TRAINING

4-51. To supervise religious support in a combat environment, the BDE UMT creates well-rehearsed coordination, synchronization, and communication techniques through training before deployment. If it does not happen effectively in a garrison/training environment then, quality religious support may not happen in a stressful and fast-paced combat operation. The responsibility for training rests heavily on the BDE UMT. Quality combat survivability training is available at the unit level. Religious support training is available at division and higher level. This does not prevent BDE Unit Ministry Teams from conducting training within the BDE. Training at this level is crucial to support the commander's mission. The BDE UMT needs to be proactive with an approved, staffed and coordinated training plan that takes into account Officer and Enlisted Critical Task lists as part of the total BDE training plan.

4-52. In addition to the BN training topics and individual and collective responsibilities, the BDE UMT must be personally involved in the training of subordinate UMTs. The BDE UMT plays a critical role in the development of the subordinate unit's junior chaplains and chaplain assistants. Success in providing spiritual support to soldiers requires adequate training. The UMTs of a BDE require sufficient training to enable it to mobilize, deploy, and provide quality spiritual support to soldiers anywhere in the world. The battlespace fixes the directions and goals of training. The BDE UMT is the training management team for subordinate UMTs. Their responsibility is to initiate, plan, and coordinate training for all UMTs in the command. The BDE UMT will make training effective by:

- Developing and communicating a clear training vision.
- Requiring subordinate UMTs to understand and perform their roles in unit, technical, and individual training.
- Being proficient on their mission essential tasks.
- Continually assessing training completed.
- Eliminating training distractions.
- Developing near and far term training plans.
- Being systematic and aggressive.

4-53. Suggested training topics include—

- Land navigation—mounted/night/navigation.
- Operations order process—
  - Developing and writing a religious support estimate.
  - Writing a religious support annex.
  - Writing religious support FRAGOS.
- War-gaming a battle as a brigade UMT combat team.
- Communications rehearsal—
  - Daily reports.
  - Communicating vital information.
  - UMT net call.
  - Practicing alternate means of communication.
  - Practicing cross communication between BN UMTs.
- Rehearsing the RSP of succession.

- Re-supply training: ordering and getting ecclesiastical supplies through logistics channels.
- Map reading, military symbology, and battle tracking.
- Conducting Religious Area and Impact Assessments
- UMT TSOP development and rehearsal.
- Chaplain assistant coordination of the RSP in absence of the Chaplain.
- Briefing the RSP.
- Staff Integration.

OPERATIONS PLANNING AND PREPARATION FOR UNIT MINISTRY TEAMS

4-54. Military operations consist of four types: offensive, defensive, stability and support. Each type consists of four phases: pre-deployment, deployment, sustainment, and redeployment. The following is a discussion of the responsibilities of company and field grade UMTs in each type and through each phase of operation.

CREATING THE TASK FORCE RELIGIOUS SUPPORT PLAN

TIMELINE OF CRITICAL EVENTS + PRIORITY OF RELIGIOUS SUPPORT = TF UMTs RSP

Timeline

4-55. The most valuable resource is time. The UMT must manage time so that the RS mission gets accomplished. A TF chaplain will not be able to see every soldier and certainly not solve every problem. Set priority of effort and allocate time to that effort. You must use the tools of time management. First, the UMT extracts all the time critical events from the OPORD and places them on a time line; for example—

- Logistics package (LOGPAC) activation and deactivation.
- Enemy recon expected.
- Friendly recon/counter-recon sent out.
- TF rehearsals/brief-backs.
- CSS rehearsals.
- Engineer movements.
- Company team movements.
- Line of departure (LD) /Be-Prepared-to-Defend times.
- Jump Aid Station (JAS) moves out, and any other time driven events.

4-56. Second, the UMT determines the intent of religious support during this phase of the battle. Knowing the religious support intent will help determine how much time to spend on visitation, where to spend it, and where priorities lay.

4-57. Third, the UMT determines soldier activity level. The Commander's Intent and unit warfighting doctrine must be understood. With these in mind, the UMT will know what tasks the company teams must accomplish. The UMT may then gauge when is the most effective time to visit and provide religious support.

4-58. Fourth, the UMT determines limiting factors like restrictions in the OPORD, ROE, METT-TC, and personal state of readiness.

4-59. The RSP is an informal plan and is flexible to meet mission requirements. It is likely to change during the execution phases of the battle because not everything goes as planned. It is easier to alter an existing plan than create (and communicate) one in a crisis.

Priority

4-60. Priority of religious support is to be identified as early as possible.

4-61. First, the UMT takes into account what has been experienced by each company in the TF, and their role in the coming mission. The UMT must determine what and where the needs are, and which unit has priority for religious support.

4-62. Second, the UMT considers morale factors known before the battle, leadership issues, human sustainment issues (food/rest/water), and interpersonal conflicts.

DEVELOPING PRIORITIES FOR THE TASK FORCE UNIT MINISTRY TEAM

4-63. The driving force behind the planning process of a TF UMT is determining its religious support priorities. Where will you place your main efforts and why? The UMT must create a time line that includes the unit's critical-time-driven-events. Working with this data, the UMT can then select blocks of time for priority religious support.

1. Study the Brigade OPORD/OPLAN and extract the following information: • The mission assigned to your battalion. • The BDE Commander's intent for this mission. • Attachments / Detachments • The RS missions assigned to the UMT - Unit and Area Support. • Channels of communication with the BDE UMT, and with UMTs in the BDE area of responsibility (AOR). • Succession plans in case the BDE UMT becomes a casualty.
2. Study the Battalion OPORD / OPLAN and extract the following information: • The missions assigned to the various Battlespace Operating Systems (BOS). – Intelligence (S2 / G2/Scouts) – Maneuver (combat companies) – Fire Support (artillery, mortars, close air support) – Air Defense (Air Defense Artillery) – Mobility & Survivability (Engineers) – Logistics (support, supply, repair, maintenance; medical) – Battle Command (Command and Staff)

3.	***Identify the main effort company(s), the supporting attack company(s), and the Scout's mission.***
4.	***Identify the first location for the medical units (BAS, JAS, FSMC), and the next several places each will jump by phase. Locate the AXPs and plot them on your map.***
5.	***Determine the priority of religious support the UMT will perform:***
	• Main effort company(s)
	• Supporting attack and reserve company(s)
	• Ordinances/ Rites / Sacraments / Worship services
	• Visitation / Counseling
	• Senior leader religious support
	• Company / Platoon leader religious support
	• Individual soldiers
	• Morale problem areas
	• Honor the dead - Memorial Ceremony/Service
	• Scouts
	• Area Support requirements
	• Religious support to other UMTs
	• Ethical / Moral counseling
	• Displaced Persons / Detained Person religious support
6.	***Create a time line for your religious support***
	• Write in the critical events of the day that effect the UMT.
	• Identify blocks of time for priority religious support.
	• Write in your priority religious support objectives in the priority religious support time blocks.
	• Consider flexible time factors in a high operation tempo (OPTEMPO) and fluid combat environment.
	• Choose an alternative plan in case conditions change unexpectedly.

WRITING RELIGIOUS SUPPORT APPENDIX

4-64. ***Consider the audience.*** The religious support appendix to the Service Support Annex is for the entire command. The religious support appendix must contain information that is both specific to subunit UMTs, and important to the whole command. The religious support appendix must support the commander's mission.

4-65. ***Only submit the appendix when it's needed.*** There are many times when there is no critical information to contribute. When appropriate, submit an appendix/matrix or a paragraph for the CSS annex. (See Appendix D)

4-66. ***Content is key, not the format.*** The UMT has no command authority so all tasks must be coordinated with commanders. Working with the S3 / G3, insert tasks for subordinate UMTs into the main body of the OPORD as tasks to subordinate units in paragraph 3. This will ensure command awareness and support for these tasks. Space for the religious support appendix varies from three pages in the BDE OPORD; one page for a matrix; or possibly only have space for a paragraph. Use the correct military format. A matrix can be utilized in keeping with the styles and formats used in the unit (See Appendix D). Or write a concise paragraph for inclusion in the CSS Annex. Ensure the content only states what is critical to support the mission.

4-67. ***A good TSOP reduces OPORD size.*** TSOP material is not repeated in the RS Appendix! The TSOP is an effective, usable, and combat oriented document. The content of the TSOP will help in the preparation of appendices and pre-combat coordination. UMTs must possess the RS portions of higher and lower units TSOPs.

4-68. ***Intent should be clearly written.*** Write the concept of religious support so higher and subordinate UMTs understand what is to be accomplished.

BRIGADE

4-69. The primary mission of the brigade UMT is to coordinate religious support for all soldiers and authorized civilians in the brigade commander's area of operations (AO). They provide religious support to soldiers in the brigade headquarters and others in the AO without religious support assets. They provide critical guidance and technical control to subordinate UMTs.

RELIGIOUS SUPPORT COVERAGE PLAN

4-70. The brigade UMT coordinates the brigade UMTs religious support throughout the brigade AO, thus ensuring that all units and individuals operating within the AO are provided religious support. This includes refugees, displaced persons, civilians in the area of operation, and enemy prisoners of war. The support of civilians and Enemy Prisoners of War is the responsibility of the commander; the commander may or may not require the chaplain to provide or perform religious support. As part of the Army component, civilian employees and contractors will be present providing essential work. Pending higher guidance issued in the division OPORD, the UMT will assess the requirements of civilian employees and contractors deployed in the AO and will include them in the religious support plan, at all levels.

CONTRACTORS

4-71. Chaplains are the staff officers in charge of coordinating religious support for contractors. All contractors who are clergy/lay ministers and readers will submit an application for approval as prescribed under Army Regulation (AR) 165-1 for supervision and coordination by military chaplains. UMTs will make every effort to plan for adequate religious support resources with their unit such as literature and sacramental supplies to support the religious needs of contractors. UMTs will plan for area religious support coverage to include contractors. Coordination for additional ecclesiastical supplies or religious support personnel will be with the supervising UMT. Chaplains will provide support for death notifications, Red Cross notifications by command, liaison with CONUS clergy, or others for contractor casualties. Religious support is coordinated through the chain of command.

RELIGIOUS SUPPORT MISSION TASK

4-72. Before the battle, the supervisory UMT submits a religious support appendix that both gives clear guidance to subordinate UMTs and allows all unit commanders to be informed of the religious support provided. The chaplain assistant gathers information that will allow the UMT to battle track the operation and current unit positions, understand overlays and SOI information, and successfully navigate the battlespace

4-73. The brigade UMT also coordinates with combat support and combat service support units in the brigade AO. Units with organic UMTs may need additional faith group support. Some UMTs may provide religious support for units whose organic support is not available. Time, distance, and battlespace conditions may restrict the travel of the UMTs and make religious support difficult.

Preparation for Combat

4-74. The brigade UMT uses the religious support estimate to plan religious support for all units in the brigade AO. It assesses the faith group needs of the battalions and requests additional support from the division UMT. The brigade UMT coordinates times and locations for forward link up points with the maneuver battalions.

4-75. The brigade UMT coordinates also with combat support and combat service support units in the brigade AO. Units with organic UMTs may need additional faith group support. Some UMTs may provide religious support for units without organic religious support or for units whose organic support is not available. Time, distance, and battlespace conditions may restrict the travel of the UMTs and make religious support difficult.

4-76. During offensive operations, the brigade support area (BSA) moves frequently. The brigade UMT may establish an alternate operational base in the BSA, but it must know how to move its equipment if the BSA relocates. As an alternative, the team may choose to establish its operational base near the brigade tactical operations center (TOC), where the majority of planning and communications are found. In the final analysis, the BDE UMT must locate in a position that best allows it to facilitate RS across the brigade. Regardless of location, the brigade UMT must be prepared to displace quickly.

The brigade UMT attends rehearsals to ensure religious support is fully integrated with the scheme of maneuver and the CSS plan.

4-77. During rehearsals, the brigade commander ensures the BSB commander has chosen adequate main supply routes (MSRs), future locations, logistics release points (LRPs), and refueling sites. The brigade UMT monitors this part of the rehearsal closely to integrate religious support with the scheme of maneuver and CSS plan.

4-78. The brigade UMT monitors the tactical situation, the status of each battalion UMT, and plans for future operations.

Develop Intelligence through reconnaissance and intelligence assets thru Exercise Command and Control

4-79. The brigade UMT monitors the tactical situation, the status of each battalion UMT, and plans for future operations. If changes in METT-TC warrant adjustments in religious support, the brigade chaplain recommends changes to the brigade commander or executive officer. Reassigning a chaplain or chaplain assistant, or changing the mission of a battalion UMT, requires a FRAGO. The brigade UMT must be prepared to augment TF UMTs in the event of mass casualties.

Perform Logistics and Combat Service Support

4-80. When the brigade reaches the objective, casualty evacuation increases. The brigade UMT again assesses the requirements for additional religious support in the AO. If the number of casualties is high, the team may choose to move to the forward support medical company in the BSA.

If the number of casualties is high, the brigade UMT may choose to move to the forward support medical company in the BSA.

4-81. If a TF UMT has not submitted a current SITREP, the brigade UMT NCOIC contacts the team to determine its status. When feasible, the brigade UMT moves forward to the TF UMTs to provide pastoral care. Priority goes to the TF executing the main attack.

4-82. If the brigade continues the mission, time may be very limited. This phase may start with reconnaissance, or it may go directly into deployment for another assault.

4-83. The brigade UMT considers the spiritual condition of the soldiers in follow-on planning. The severity of the previous operation and the number of casualties suffered are key factors in planning.

4-84. Split Operations could include a combined Legal, Medical, UMT and Civil Affairs Cell as an active battle-tracking proponent in TOC operations, while a separate religious support cell, employing all BSA UMTs, operates in the BSA TOC. This type of operation promotes better use of RS assets, prepares the BSA Chaplain to assume the responsibility of the TF Chaplain in the event of loss, provides greater opportunities to minister to TF BOS systems, and produces two avenues for receiving TF UMT daily reports. The TF UMT maintains contact with the BSA cell through digital nonsecure voice telephone (DNVT) channels.

4-85. If the brigade is operating as part of a division mission, the brigade UMT of the follow-and-support brigade establishes liaison with the UMT of the lead brigade to facilitate pastoral care to casualties and other religious support. The follow-and-support brigade may provide evacuation and treatment for casualties of the lead brigade, so that the lead brigade can continue its tempo. This mission is of particular concern to the brigade UMT because of its complexity, intensity, and momentum.

Division Cavalry Squadron

4-86. The division cavalry squadron may operate across the entire division front or along one or both flanks. When the squadron operates forward of the brigades, the squadron UMT coordinates with the brigade UMTs for the care of casualties and for additional religious support.

DEPLOYMENT

4-87. BDE UMTs are prepared at all times to deploy with their respective unit. Their deployment plan is synchronized and coordinated with the Brigade command and staff, subordinate TF UMTs, and higher supervisory UMTs. The UMT trains throughout the year on the needed skills and tasks for deployment. Planning considerations include—

- Air/Ground flow schedule.
- Intelligence information.
- Subordinate UMT readiness.
- Unit spiritual fitness.
- Ecclesiastical resupply kits and other religious support material.

Pre-combat Planning Considerations

4-88. After completing the planning process, the UMT is not yet ready to begin religious support operations. Coordination with key people and pre-combat checks still need to be completed. Listed are examples of areas for coordination, inspection, and team member responsibility.

- Chaplain—
  - Worship/Counseling preparation.
  - Coordination with—
    - Executive Officer for religious support plan/appendix; staff meetings; daily activities.
    - Commander for his guidance and focus; give advice on spiritual fitness of command.
- Chaplain Assistant—
  - Coordination with headquarters and headquarters company (HHC)/headquarters and headquarters troop (HHT) commander for bivouac site; local security plans; food and fuel; planned moves of trains or TOC.
  - Checks and Activities—
    - Individual precombat inspection (PCI).
    - Weapon/ammo.
    - Mission oriented protective posture (MOPP) gear.
    - Survival gear.
    - Food.
    - Water.
    - Issue equipment.
  - Vehicle PCI—
    - Meals ready to eat (MREs) stored.
    - Operator's vehicle material (OVM) stored.
    - Fuel.
    - Water.
    - PMCS complete.
    - Radio and frequencies, crypto, fill.
  - Team PCI—
    - Map with overlays.
    - CEOI/SOI information.
    - Night Vision Goggles with spare batteries.
    - TSOP.
    - Compass.

      - M256 kit.
    - Movement Plan—
      - Map Recon.
      - Plan posted in Combat Trains Command Post (CTCP)/Field Trains Command Post (FTCP)/Tactical Operations Center (TOC).
    - Ecclesiastical Supplies—
      - 30/60/90 day supply.
      - Chaplain's Kit complete.
- Unit Ministry Team—
    - With subordinate unit commanders/CSMs/1SGs—
      - Plan subordinate unit worship services.
      - Identify times and locations for visiting soldiers.
      - Identify special spiritual needs of the command.
      - Identify other special needs and issues.
    - Higher and lower-level ministry teams—
      - Religious Support Plan (at least two levels up).
      - Location during phases of battle.
      - Religious support requirements.
      - Availability of and plans for denominational support.
      - Emergency ecclesiastical re-supply.
      - UMT Replacements.
    - Staff—
      - Casualty information.
      - Intelligence updates (critical for the UMT to keep themselves constantly updated).
      - METT-TC updates.
      - Medical Plan updates.

OPERATIONS

Staff Coordination

4-89. Before departing the operational base, the UMT leaves a copy of its travel plan in the command post (CP) or TOC. The chaplain coordinates with the S1/S4 representative to receive messages and requests for religious support. The brigade chaplain is normally expected to attend shift change and battle update briefings (BUB). The brigade chaplain should ensure a means whereby those who try to contact the UMT can leave messages and that the UMT location is known at all times.

4-90. During offensive operations, the brigade support area (BSA) moves frequently. The brigade UMT may establish an operational base at the brigade TOC. The BDE TOC provides the UMT—

- Current operational updates.
- Availability to the commander and key staff personnel.
- Battle tracking.

4-91. As an alternative, the team may choose to establish its operational base in the BSA. The TOC is preferable. Regardless of location, the brigade UMT must be prepared to displace quickly.

4-92. During rehearsals, the brigade commander ensures that the Brigade Support Battalion (BSB) commander has chosen adequate MSRs, future locations, re-supply points, and refueling sites. The brigade UMT monitors this part of the rehearsal closely to integrate religious support within the scheme of maneuver and CSS plan.

Note: When moving independently among task force and slice units, the brigade UMT must be aware of the potential for being left behind. To prevent this situation, UMTs must have dedicated communications equipment that is compatible with the unit's secure communication equipment.

Coordination with the CSS planners
1. Refer to FM 71-2, CSS chapter.
2. The BDE will issue its OPORD to the TF leadership early. The BDE OPORD contains the BDE UMT's Religious Support Appendix. Coordinate with your S4, and ensure his familiarity with the religious support appendix.
3. The TF Cdr will issue a warning order to the TF. A possible format is in FM 71-2, appendix
4. At this point one of two things might happen; either the S4 will return to the CTCP and convene a meeting with the battalion maintenance officer (BMO), medical platoon leader, S1, and the Chaplain to review the BDE order and receive feedback for CSS planning; or the S4 stays at the TOC and holds the CSS planning meeting there. If the Chaplain is occupied with ministry, the chaplain assistant attends the CSS planning meeting and provides UMT input.
5. The Cdr will issue the OPORD to the command at the TOC or other designated location. One member of the UMT should attend that meeting.
6. After the UMT has studied the OPORD, coordinate with the medical platoon leader. Find out first hand what the CSS plan is, especially the planned locations during each phase of the battle.
7. Attend the CSS rehearsal and listen for changes in the plan. During the rehearsal, let everyone know where the UMT will be during each phase of the battle.
8. Monitor the Admin/Log net to stay current. Often the Command puts out changes to the plan on short notice. The tactical situation changes rapidly and you might find the planned attack has changed into a hasty defense. Be alert, your life depends on it.

SITUATIONAL AWARENESS

4-93. Frequently the operation does not take place as planned. The UMT must know how to get information from a number of sources in the unit. There will be times when the back-channel information flow is more timely and accurate than the official flow.

4-94. Timely coordination with the CTCP leadership (S4 and S1) and their staffs before deployment makes it easier to coordinate and publish UMT FRAGOs. If the CTCP has not planned to track the UMT before it goes to the field, then in all likelihood it will not track the UMTs movements and priorities when in the field. Prior coordination makes it happen.

4-95. If the UMT announces that it will visit a particular company or section, then it needs to follow through on that plan. The UMT must carefully consider any information it places in the TF OPORD.

4-96. The Chaplain is a key player on the CSS staff. There are critical missions the chaplain must accomplish. The UMT must integrate with the CSS staff before deployment and understand all the forms, reports, and planning process used by the staff. The TF XO is responsible for all CSS within the TF and the chaplain must insure the XO is familiar with the critical religious support coordination factors that must be completed before deployment.

4-97. Finally, the CSS staff will produce estimates of anticipated casualties and expenditures for the upcoming battle. These estimates help the UMT to plan their religious support mission more effectively.

THE BRIGADE SUPPORT BATTALION UNIT MINISTRY TEAM IN THE BRIGADE SUPPORT AREA

4-98. The Brigade Support Area (BSA) is generally located behind the rear boundary of the TF sector. The BSA contains the support elements for the entire Brigade. The Engineer Sapper Battalion, the Brigade Support Battalion (BSB), and the field trains for all the task forces are just a few of the CSS assets in the BSA (FM 71-3). Normally, the TF UMT rarely visits this area because of the distance from the TF fight. There will be a supply route (SR) from the BSA into each TF sector. Each BDE and TF Chaplain should read the CSS chapter in FM 71-2 to gain an understanding of the composition and doctrine of the Field Trains in a BDE.

4-99. The HHC Commander is in charge of the Field Trains and will have a Field Trains Command Post (FTCP). The FTCP has the job of tracking the battle and coordinating the CSS actions in the BSA. The support platoon locates here (including mess and maintenance) and thus the LOGPAC system for the TF. The HHC Commander will also attend the "tenant's meetings" given by the BSB leadership to coordinate CSS issues.

4-100. Communication with the FTCP is difficult due to distance. The best method is to determine how the S4 communicates with the FTCP and use the same method. Possible pathways of communication are: the mobile subscriber equipment (MSE); frequency modulation (FM) radio communications with the TF A/L (Administrative / Logistics) net are possible if the distance is not too great. Sometimes a messenger can be used. Individuals to carry messengers can be found on the LOGPAC convoy. Link up with the LOGPAC convoy at the Logistics Release Point (RLP) in the TF and meet with the Support Platoon Leader or Platoon Sergeant.

4-101. The FTCP Command and Staff must understand that the BDE UMT or another UMT might work with them on various issues. For example, the BSB UMT might have the mission of working with the replacements going forward before their assignment to the individual companies. The BDE UMT might use the FTCP to pass messages/reports forward to the TF UMT and vice versa. One of the UMTs in the BSA normally provides religious support for the soldiers in the field trains.

4-102. The BSB is a critical asset for the BDE. The medical company will provide evacuation assets to the TF medical platoon as well as serve as a replacement treatment team if needed. The flow of ambulances between the BSB medical company and the TF BAS provides a readymade transportation/communications network. Companies evacuate wounded to a Patient Collection Point where an ambulance from the BAS/JAS collect the wounded. The wounded receive emergency treatment at the BAS/ JAS. A BSB medical ambulance will then evacuate them directly or through an ambulance exchange point (AXP). The soldier receives further treatment in the BSB before further evacuation. The BDE chaplain will assign a UMT to work with the wounded in the BSB (normally the BSB UMT). Communications can go by the evacuation route. If the TF UMT uses this method, make prior coordination so BSB ambulance personnel understand their role in passing messages, and the BDE UMT arranges to get the messages. This should be in the TSOP for both the TF and the BSB to insure no misunderstanding. Again, the TSOP is very important!

4-103. Tenant's meetings held at the BSA by the BSB leadership are for all elements of the BSA community. A representative from the BDE S1/S4 command post will normally attend as well as the TF HHC Commander from the TF Field Trains Command Post. The BDE UMT normally has its base of operation in the BDE S1/S4 command post area. Information and messages may go with the TF HHC Commander to the BDE S1/S4 representative.

THE ENGINEER BATTALION

4-104. The Engineer UMT is usually based in the BSA. Engineer activities are directed through the engineer Administrative/Logistics Operations Center (ALOC) which is located near the BDE TOC. The Engineer UMT should assume some of the area RS requirements in the BSA. Normally, the engineer assets in the TF are in known locations at a Class IV point. If the Engineer UMT elects to move forward into the TF sector to visit its scattered unit, it should first coordinate with the BDE UMT.

4-105. Religious support within the BSA and for the entire BDE is critical. The BSB UMT should have a definite RSP for their area of operations and for the BDE through the BDE UMT. The BDE UMT should provide specific guidance and tasks for the BSB UMT. Some tasks are as follows—

- Provide religious support to the medical staff.
- Read casualty feeder cards to determine if religious support has been given by forward TF UMTs.
- Receive/send messages from/to TF UMTs then relay messages to BDE UMT.
- Brief and prepare replacements for combat operations.
- Provide BDE UMT information on soldier replacement issues, medical trends, number and types of casualties, and the BSA situation.

RELIGIOUS SUPPORT TO CASUALTIES

4-106. Spiritual care to wounded soldiers is one of the core functions of the UMT. It contributes to the soldiers' well-being and recovery. Prayers, scripture readings, rites, ceremonies, and sacraments constitute religious support for wounded or dying soldiers. While the chaplain gives care to individual casualties, the chaplain assistant moves among the other soldiers to provide care and to identify those in greatest need of assistance.

4-107. During combat operations, maneuver battalion UMTs may locate with the battalion aid station. If the medical platoon divides the aid station into two or more treatment teams, the UMT should consider locating where the most casualties are expected. Based upon experience, the UMT may split for a short time to maintain contact with the separate treatment teams. However, this is based upon training, experience, and necessity. The chaplain must carefully weight all aspects.

4-108. The UMT considers the physical condition of the casualties and gives priority to the most seriously injured. The following medical categories help the team determine priorities for care:

- ***Expectant:*** casualties with very serious injuries and low chance of survival. The UMT provides pastoral and spiritual care for the dying as its first priority.
- ***Immediate:*** casualties with a high chance of survival if medical measures are accomplished quickly.
- ***Delayed:*** casualties needing major surgery, but who are able to tolerate a delay in surgical treatment.
- ***Minimal:*** casualties with relatively minor injuries that can be treated with self-care. Do not ignore these casualties as most will be treated and returned to combat with potentially high levels of anxiety.

RELIGIOUS SUPPORT TO CAREGIVERS

4-109. Religious support to caregivers is a vital part of the UMT's mission. Often the gravity of the situation, the need to give aid quickly, and the absorbing engagement of the moment will carry caregivers through the crisis. When the crisis is concluded, however, the impact of what has happened may strike the caregivers suddenly and dramatically. Medical personnel and mortuary affairs personnel may be in particular need of religious support at the conclusion of a traumatic event.

RELIGIOUS SUPPORT TO GRIEVING UNIT MEMBERS

4-110. Religious support to everyone involved in a military operation is critical. Often survivors are stunned and immobilized by the loss of fellow soldiers. The emotional and spiritual impact of loss requires the finest skills of the UMT to deal with the denial, anger, guilt, and fear, and to facilitate the continuation of life and service.

4-111. Care should begin as soon as possible after the loss occurs. The knowledge that soldiers have received religious support gives comfort and encourages grieving soldiers to move on with their lives. Following the immediate intervention, supportive follow-on care should be provided when soldiers re-deploy to their home station.

OFFENSE

4-112. The offense is the decisive form of combat operations. The main purpose of the offense is to defeat, destroy, or neutralize the enemy force. While strategic, operational, or tactical considerations may require defending for a time, defeating the enemy requires shifting to the offense. Rapid movement characterizes offensive operations over large areas. The types of offensive operations are movement to contact, attack, exploitation, and pursuit.

4-113. Offensive operations place great demands on UMTs. The constant movement of units and troops, and the operation tempo (OPTEMPO) of the battle limit the UMTs' access to soldiers. The religious support focus will vary from mission to mission. Once an offensive operation begins, UMTs in combat units will provide ministry on a different scale and at different times from those in combat support and combat service support units.

OFFENSIVE PLANNING CONSIDERATIONS

4-114. The UMT plans religious support for each phase of an operation. Offense planning considerations include—

- Develop Intelligence through reconnaissance and intelligence assets.
- Conduct Maneuver.
- Deployment.
- Employ Firepower.
- Protect the Force
- Exercise Command and Control (C2)
- Perform Logistics and CSS Consolidation and Reorganization or Continuation.

(See FM 100-40 for more detail.)

INFANTRY OR ARMOR BATTALION/TASK FORCE

Preparation for Combat

4-115. The Task Force (TF) UMT begins planning for religious support by considering METT-TC. The mission and the time available will determine how the UMT provides religious support. With little time, the team plans and prepares quickly not to miss opportunities for religious support to soldiers. For example, the time available for religious support before a hasty attack is much less than the time available for religious support before a movement to contact or a deliberate attack.

4-116. The UMT carefully establishes religious support priorities. It considers the order of march, the units conducting the main and supporting attacks, and the severity and number of casualties in previous missions. Priority is given to elements that have sustained the most casualties or to those that will be engaged at the earliest. Pastoral care to casualties is always the priority. Therefore, the UMT coordinates with the medical platoon when casualty evacuation plans are formulated.

Develop Intelligence through Reconnaissance and Intelligence assets.

4-117. This phase begins as soon as possible after the unit receives a mission. Reconnaissance elements seek to detect enemy activity and the physical characteristics of areas of interest to the commander.

> The scout platoon is an early priority for religious support.

4-118. The TF scout platoon begins its reconnaissance and surveillance as soon as the mission is received from the brigade. It may leave as much as 24 hours before the main body crosses the LD. The scout platoon is an early priority for religious support.

4-119. While the scout platoon conducts the reconnaissance, the rest of the unit continues preparation for combat. This preparation normally takes place in a tactical assembly area (TAA). Worship services, sacraments and ordinances, individual counseling, and visitation are possible and very important for soldiers during this phase. The UMT coordinates with company commanders for times and places for religious activities. An excellent opportunity for this coordination is the meeting at the LRP.

> The UMT coordinates with company commanders and first sergeants for time and place of religious activities.

4-120. The UMT utilizes the intelligence preparation of the battlefield (IPB) process by the S-2 staff section to identify any aspect within the AO that will affect the manner in which they accomplish the religious support mission.

Conduct Maneuver

4-121. When attacking from positions not in contact with the enemy, units normally stage in a TAA. The TF does not move all at once. While the line companies move toward the LD, the main command post, the combat trains, battalion aid station (BAS), support elements of the TOC, mortars, and other elements of the TF prepare for the operation. Religious support is given to each element of the TF before it begins movement. Soldiers who are preparing for the operation may not have time to gather for worship.

> The Task Force UMT normally travels with the combat trains or with the BAS.

4-122. The TF UMT normally travels with the combat trains or with the BAS. If the team is visiting soldiers in other locations before movement, it must determine ahead of time which routes will be clear of obstacles, how long it will take to return to the BAS, and the location of link up points. The UMT times its movement to precede or coincide with the scheduled start time of the BAS. All movement is coordinated with the S4.

4-123. If the medical platoon intends to split the BAS, the team must plan support for more than one site. The UMT must know where the TF surgeon will locate; where most of the casualties are expected; and, if the UMT needs to separate, how it will maintain two operations.

4-124. The displacement of critical CSS elements, such as the combat trains, BAS, Unit Maintenance Collection Point (UMCP), Logistic Release Points (LRPs), and Casualty Collection Points (CCPs) will be triggered by the TF crossing phase lines or by other control measures. This means of movement allows the CSS elements to remain responsive to the task force. The TF UMT monitors the tactical situation to know when the TF is approaching each phase line and thus when it may need to move or shift its religious support focus.

4-125. Because of distances and the OPTEMPO, most units will re-supply on the move. The TF UMT will plan accordingly for its own re-supply.

4-126. As the TF moves, communication becomes difficult. The UMT will find it hard to monitor the tactical situation. Good planning, rehearsals, and the ability to adapt on the battlespace enable the UMT to provide comprehensive religious support despite the confusion and chaos of battle. Coordination with key staff members and the executive officer is critical.

The UMT must employ sleep plans and staggered shifts when necessary to remain fully capable of night operations.

4-127. Task force operations are continuous. The UMT must be fully capable of night operations, employing sleep plans and staggered shifts when necessary. If the TF moves during the hours of darkness, the TF UMT must maintain contact with other elements with which it is traveling.

4-128. The attacking force moves to positions of advantage during this time. The basic forms of maneuver are envelopment, penetration, frontal attack, turning movement, and infiltration.

4-129. In the maneuver phase, religious support may be restricted to small numbers of soldiers during halts in the movement, or at casualty collection points along the route of march.

4-130. During the assault, casualties are likely to increase. The TF UMT will be heavily engaged with providing religious support to casualties and caregivers. If overwhelmed with casualties, the BAS declares a MASCAL situation. The TF UMT requests additional support during a MASCAL from the brigade UMT.

4-131. During a MASCAL the TF UMT requests additional support from the brigade unit ministry team.

Employ Firepower

4-132. The fire support coordinator (FSCOORD) integrates fire support into the unit's maneuver scheme. Religious support during this phase is similar to the maneuver operation.

Protect the Force

4-133. To protect the force a commander can employ any number of actions to include: maintaining a high tempo of operations; conducting security operations, which include organizing and employing covering, guard, or screen forces; conduct area security operations; employ operations security (OPSEC) procedures; execute deception operations; conduct defensive information operations; employ camouflage, cover, and concealment, and conduct active and passive air defense operations.

4-134. The UMTs movement in the battlespace, communication methods, and information management techniques should reflect the commander's security and force protection objectives. The UMT closely monitors force security issues that may affect religious support planning The UMT aggressively monitors the battle and the commander's use of command and control (C2) in allocating forces. The decision-making process and the format for plans and orders are found in FM 101-5. Below are the UMTs planning considerations within this staff process.

The Goal: Comprehensive/Synchronized Religious Support		
Staff Action	***UMT Action for input to staff***	***UMT Action for internal use***
Receive the Mission	-Mission Analysis: Can we support the mission and how? -Coordinate with S4 and S1. -Coordinate with higher UMTs.	-Begin rehearsals of known actions/drills. -Execute current RS plan. -Begin considering the unit's proposed COA's for support of the mission.
Conduct Estimates by each staff section/	-Prepare RS Estimate of current/projected unit condition. -Provide input to staff on AOR religions and culture.	-Communicate issues to subunit UMTs. -Continue Spiritual Care.
Restated mission to Cdr/ Receive Initial Cdr's Guidance	-Review religious, moral, and ethical issues.	-Continue prep for next mission.
Develop/compare/ recommend the 'best' COA to Cdr	-Input to COA's for tactical mission. -RS Implications of each COA. -Compare feasibility/ acceptability/ supportability of each COA that the staff produces.	-Further develop RS plan for support of the upcoming mission. -How will we synchronize RS? -Continue provide RS to unit and area personnel.
Receive Cdr's Estimate including DECISION and Cdr's CONCEPT	-Refine RS concept to comply with Cdr's concept.	-Communicate with subunit UMTs concerning their own preparation of subunit RSP needs and issues.
Prepare Plans and Orders	-Submit RS annex/paragraph if needed.	-Update RS Section NCOIC. -Continue RS.
Receive Cdr's approval of Plans and Orders	-Pray with Cdr, and give 1 min update of RS situation.	-Continue information gathering. -Continue RS.
Issue Plans and Orders	-Attend OPLAN meeting; be prepared to brief.	-Communicate RS plan and provide support to higher RMSTs and subunit UMTs.
Rehearsals and Troop Leading Procedures	-Attend rehearsals, insure RSP is integrated and understood.	-Aggressively execute RS plan.
Conduct mission	-Begin planning for follow on mission, sequels, branches.	-Provide prioritized RS. -Battle track subunit UMTs.

4-135. The UMT must remain flexible and support changes to the plan due to METT-TC considerations and the commander's priorities.

PERFORM LOGISTICS AND COMBAT SERVICE SUPPORT

4-136. On reaching the objective, the unit eliminates remaining resistance, consolidates its gains, and reorganizes as needed. The unit may immediately continue the attack to a follow-on objective, or it may hold in place and set up a hasty defense.

> The UMT provides religious support to the force paying attention to leaders and those who show signs of battle stress.

4-137. The time the TF spends during this part of the offensive operation will depend upon subsequent missions. If there is sufficient time, the UMT moves forward to maneuver elements. The UMT provides religious support to the force paying special attention to leaders and those who show signs of battle stress. The team may also conduct memorial services or ceremonies as directed by command.

4-138. A successful assault may develop into an exploitation or pursuit. The effectiveness of an exploitation or pursuit depends on rapid transition and vigorous execution. The UMT must be prepared to make the transition quickly with the least disruption to its operations.

TRAVEL IN THE BATTLESPACE

4-139. A UMT that travels to forward elements by itself is at great risk. Travel with the TF LOGPAC convoy reduces the security risk and the chances for getting lost. The LOGPAC moves from the field trains along the MSR to an LRP where the company first sergeants take control of their company LOGPACs.

> The UMT meets with company first sergeants at the LRP.

4-140. Based on the scheme of maneuver, the TF S4 selects LRPs well forward. For security reasons, they are only used for short periods of time. The UMT finds LRP times and locations in the OPORD.

4-141. The S4 and company first sergeants meet at the LRP to discuss future logistical requirements. At this meeting, the UMT discusses the religious support needs of soldiers and coordinates for services and visits to the line companies with the first sergeants. The UMT then may go to one of the companies with the first sergeant and company LOGPAC

4-142. When the company has completed its re-supply, the first sergeant returns his portion of the LOGPAC convoy to the LRP. The TF support platoon leader reforms the convoy and leads it back to the field trains. The TF UMT follows the first sergeant to the LRP and returns to the combat trains or, if time permits, travels to the field trains.

4-143. Engineers. The engineers provide an alternative to traveling with the LOGPAC. Because LOGPACs are often conducted during times of limited visibility, the UMT may find travel with elements of the engineer company more effective. The engineers travel to all the company locations to dig in hasty defensive positions, clear mines and obstacles, and prepare for the follow-on mission. The UMT coordinates with the S4 to link up with the engineers for travel to the companies.

RELIGIOUS SUPPORT OF DEFENSIVE OPERATIONS

4-144. Army forces conduct defensive operations as part of major operations and campaigns in combination with offensive operations. The basic task of the defense is to force the enemy to attack under unfavorable circumstances. Military forces defend only until they gain sufficient strength to attack. Though the outcome of decisive combat derives from offensive operations, it is often necessary, even advisable, to defend.

4-145. Defensive planning considerations are very similar to offensive considerations. In the defense, the UMT has more time to synchronize the religious support plan and apply maximum religious support assets to critical areas. Shifting religious support assets as the operation evolves will require prior coordination with commanders and UMTs. Prior planning regarding the synchronization of the religious support mission and plan during different phases increases the effectiveness of the religious support defense plan.

BATTALION/ TASK FORCE

Develop Intelligence

4-146. As soldiers prepare defensive positions, the TF executes the counter-reconnaissance mission. This mission is meant to deny the enemy information about friendly dispositions and friendly preparations for battle. The counter-recon force typically consists of the scout platoon, ground surveillance radars, and maneuver elements.

> Starting as far forward as possible, while minimizing risk to exposing soldiers to artillery fire, the UMT makes brief visits to soldiers in fighting positions.

4-147. Throughout this phase soldiers are digging in and preparing fighting positions for the coming enemy attack. The UMT keeps movement to a minimum to avoid disclosing fighting positions, paying close attention to cover and concealment. Starting, as far forward as possible, the team makes brief visits to soldiers in fighting positions, working rearward towards the combat trains. The UMT coordinates its travel with the TOC and the combat trains command post (CTCP).

> The UMT must carefully plan movement and religious support to avoid being caught in the confusion.

Deploy/Conduct Maneuver

4-148. The rearward passage of lines by the covering force causes considerable movement and confusion. The UMT must be cautious when planning movement and religious support to avoid being caught in the confusion. The UMT locates with the battalion aid station or LOGPAC, communicates its location to the TOC and CTCP, and deliberately avoids the passage lanes.

Employ Firepower

4-149. Enemy artillery poses a significant threat to the safety of the UMT. The UMT must be near adequate overhead cover for protection. Gathering soldiers for worship may involve undue risk.

4-150. If the enemy uses chemical or biological agents, the team will make adjustments in religious support. The team may need to postpone religious activities until the hazard is eliminated or reduced to a safe operating level. Several factors that the UMT considers include:

- Degree of engagement with enemy forces.
- Level of exposure and risk to soldiers and equipment.
- Priorities for religious support.
- Anticipated operations in response to enemy attack.

The chaplain offers sacraments and ordinances to soldiers before combat.

4-151. Sacraments and ordinances are not celebrated in contaminated areas. The UMT does not resume services until soldiers have moved to safe areas. Therefore, the chaplain takes advantage of every opportunity to administer the sacraments before soldiers engage in combat, or before soldiers are exposed to a contaminated environment.

4-152. The UMT carefully coordinates movement around and through obstacles set in place for the defense. Obstacles installed after the team has departed its base of operations can put the team at risk. The team should know the breach marking system and MSR. To avoid obstacles, it carries a copy of the obstacle overlay.

The team carries a copy of the obstacle overlay when traveling.

4-153. The UMT must be prepared to adapt to the changing tactical situation of a mobile defense that combines offensive, defensive, and retrograde actions. This combination results in a nonlinear front to create confusion for the attacking forces.

4-154. As the enemy begins the attack, the UMT shifts its attention to soldiers in the battalion aid station (BAS). From this point on, the UMT's main effort will be the care of dying and other casualties as they are brought to the BAS.

The UMT must be prepared to move on a moment's notice.

4-155. During the enemy attack, the BAS may move frequently, evacuation routes may be blocked by an enemy penetration, and defending elements may become temporarily encircled or bypassed by enemy forces. Rapidly moving enemy units may threaten or overrun the BAS. The UMT must be prepared to move on a moment's notice. Equipment and supplies not required immediately should be stowed in the team's vehicle. The vehicle should be parked in a manner that allows for rapid departure if the BAS is threatened.

Perform Logistics and Combat Service Support

4-156. As the TF consolidates and reorganizes after the battle, the companies evacuate the remaining casualties to the aid station. Once the UMT has ministered to the dying and other casualties, it assesses METT- TC to determine its next actions. If the unit is planning a counterattack, the UMT prepares to move with the unit. If not, the team moves forward to units on the forward line of own troops (FLOT) to visit soldiers. The UMT NCOIC ensures the team's supplies are adequate for follow-on operations.

Coordinate with Command

4-157. The UMT must have a clear understanding of the battlespace situation so they can prioritize tasks at the right time and location. The UMT takes advantage of the war-gaming that takes place in the MDMP to derive their decision points. The UMT coordinates with the commanders and other UMTs in the AO to ensure all religious support needs are met according to the operational plans.

BRIGADE

4-158. In the mobile defense, battalions engaged in offensive, defensive, and retrograde actions present a challenge for the brigade UMT. The UMT plans religious support for each of the units based on METT- TC. The area defense differs from the mobile defense in that all units are engaged in defensive operations. Within the defense planning considerations by the staff, the brigade UMTs responsibilities to the BDE UMTs include—

- Faith Group support.
- Operational information.
- Logistics of Faith Group specific supplies.
- Personal spiritual support to UMTs.

STABILITY OPERATIONS

4-159. Conducting religious support in stability operations is essentially the same as in offense, defense, and support operations.

Battalion

4-160. The UMT will need to reprioritize planning considerations in order to fully support this type of operation. The UMT planning emphasis will include—

- Joint and multinational religious support cooperation and communication.
- More planning for faith-based NGO/IO liaison tasks.
- Working more closely with the CMO personnel in mutual support tasks.
- Host nation religious leadership liaison responsibilities.

Brigade

4-161. The BDE or equivalent will be challenged with coordination and information management requirements. The information flow will require the proper assessment of information to allow for timely and adequate religious support. Planning considerations in addition to BN include—

- Civil Military Operations Center (CMOC) representation involvement.
- Staff integration needs.
- Media and public affairs command guidance.
- Remote site religious support coverage planning.
- Force protection.
- Religious support to Contractors.

SUPPORT OPERATIONS

4-162. The tasks for planning religious support for Support Operations are identical for offense, defense and stability operations. The support operation requires the UMT to know the restrictions, limitations, and proper responsibilities of the UMT in providing religious support in a support operation environment.

Battalion

4-163. The UMT will be directly involved with civilians and the trauma associated with the operation. This will have a stressful effect on the UMT members. The BN UMT must be prepared for the area of operation METT-TC considerations. Planning includes—

- Personal burnout and combat stress.
- A daily log (See DA Form 1594).
- Soldier and civilian debriefings.
- Continual spiritual and morale assessment of individuals in the unit.

Brigade

4-164. The BDE UMT will work closely with state agencies and joint chains of command. The religious support plan and operation must operate within the mission, concept of operation, and commander's intent. Legal restraints may hinder the free exercise of religion as it relates to the military providing religious support to civilians who request assistance and guidance. The UMTs battle book and TSOP must reflect the planning considerations. Planning considerations include—

- Mass Casualty training.
- Information on civilian and state agencies for liaison tasks.
- Critical Event Debriefing as a norm.
- Moral and ethical dimensions.

TERMINATION AND POST-CONFLICT OPERATIONS

4-165. Planning for termination of operations must be ongoing during all phases of COA development, deployment of forces, and execution of operations. It must be fully integrated within the orders process and synchronized and coordinated with the staff and UMTs. In most cases, operations will be terminated in their own right when stated objectives have been met. In some operations (e.g. stability and support), the command may have to recognize that the mission is not achievable without restructuring and/or commitment of additional assets. Termination of the mission may be the best COA. The following checklist may assist in termination planning:

- What is the planned timeline for termination?
- Will there be follow-on forces and who?
- What are the post-conflict requirements?
- Is there a force protection issue for safely withdrawing forces?
- What are the requirements for redeployment?
- What are the considerations for release of the Reserve Component forces?

- Will there be humanitarian, governmental, and infrastructure assistance required for post conflict activities? Is there coordination required?

(See JP 5-00.2 and JP 1-05 for more detail.)

TRANSITION

4-166. As with planning for conflict termination, planning for transition must extend throughout the planning process and into operations and redeployment. Transferring control of command and staff responsibilities to another organization requires detailed planning and execution. It is dependent on the situation and each transfer will have it's own characteristics and requirements. Transition planning by the UMT continually requires close coordination with higher and communication with the UMTs of the transitioning organization. Regeneration and reconstitution is a primary consideration in the transition plan. Transition planning by BDE should encompass the following:

- What is the new mission?
- What U.S. forces, equipment, and/or supplies will remain behind?
- Who will support UMTs that remain behind?
- What is the classification of information to be shared with incoming UMTs?
- What will be the new ROE?
- What are the building/Chapels requirements?
- What is the timeline for transfer?

The mission may dictate that additional planning may considerations need to be planned and coordinated for as well.

REDEPLOYMENT AND RECONSTITUTION

4-167. Reconstitution (see FM 100-9) is a method of replenishing units that have been decimated to the extent they are no longer combat-effective. Such units are withdrawn from contact to a rear area. Personnel and equipment as well as rest and recuperation are the focus. These units require special religious support. BDE UMTs must be proactive in supporting reconstitution not only for unit personnel, but the UMTs of that unit, its leadership, and caregivers as well. Religious support will encompass the following:

- Honoring the dead with memorial services and ceremonies.
- Nurture of the living with counseling, worship, and critical event debriefings.

4-168. Post-conflict requirements and impact have a direct bearing on the redeployment religious support mission. The impact of battle and the stress associated with personal loss, fear, physical and emotional fatigue demands religious support application. METT-TC is still utilized in this stage by UMTs as it was in deployment. The balancing of available assets is essential during this phase to ensure effective religious support. UMTs must factor in redeployment not only back to home station, but also to other missions. Both require extensive religious support planning. UMTs of the BN, BDE, and equivalents must not re-deploy prior to their completing their religious support mission to their respective units and other associated religious support requirements of the TF. Family support and other religious support missions at home station is the responsibility of the installation religious support mission. Planning considerations for both echelons include—

- After Action Report.
- Reunion Briefings program.
- Logistics and time table.
- Religious support coverage plan.
- Echelon of UMTs back to home station ensuring AO religious support coverage.

GARRISON OPERATIONS

4-169. UMTs participate in the religious support program of the installation in accordance with their orders of assignment by Headquarters Department of the Army. They coordinate with their chain of command and the installation chaplain for garrison religious support needs for their unit soldiers and families while non-deployed and during all phases of deployment. Planning considerations include—

- Family Readiness Group (FRG): Support by briefings/classes on family issues.
- Killed in action (KIA)/missing in action (MIA)/wounded in action (WIA) memorial ceremonies and religious services.
- Hospital visitation.
- Casualty notification support.
- Worship support.

OTHER BRIGADE AND BELOW UNITS AND SPECIAL RELIGIOUS SUPPORT CONSIDERATIONS

Special Forces

4-170. Special Forces (SF) combine at the lowest tactical level, the functions performed by conventional branches of the Army. United States Army Special Operations Command (Airborne) (USASOC (A)) is the Army component of the joint and unified command, United States Special Operations Command (USSOCOM). USASOC (A) is comprised of Special Forces, Rangers, a Special Operations Aviation Regiment, the Special Operations Support Command and Civil Affairs and Psychological Operations forces. Special Forces is commanded by the U.S Army Special Forces Command (USASFC), which has seven subordinate commands known as Special Forces Groups consisting of three Special Forces (SF) battalions each. USASFC (A) has five primary missions and related additional capabilities.

Primary Missions

4-171. The five primary missions are Unconventional Warfare (UW), Foreign Internal Defense (FID), Special Reconnaissance (SR), Direct Action (DA), and Counter Terrorism (CT). An SF unit can have elements deployed to several different areas within its assigned region. These elements range in size from the individual soldier to units of 12 to 150 soldiers. SF missions are directly influenced by politico-military considerations. A change in national security policy or strategy may change SF missions.

Additional Capabilities

4-172. Special Forces units are trained to conduct MOOTW missions. Language capabilities, area orientation, small unit, and multi-agency experience are invaluable in MOOTW. Additional capabilities include noncombatant evacuation operations (NEOs), arms control, support to domestic authorities, humanitarian assistance (HA), disaster relief, security assistance, nation assistance, counter-drug operations, combating terrorism, peacekeeping and peace enforcement, show of force, support to insurgencies and counterinsurgencies, personnel recovery, and special activities.

4-173. The Special Forces branch or related MOS is the only non-accessioning branch or MOS. A soldier selected for the Special Forces branch or MOS has successfully completed service in an accession branch or MOS and graduated from the Special Forces Qualification Course.

UMT Personnel Qualifications

4-174. Religious support in the Special Forces community requires unique qualifications. The most critical is the ability to provide sustained, independent religious support. Others include—

- Security clearance for all UMT members equal to the mission classification.
- Airborne qualification.
- Extensive orientation in special operations, or graduation from the Special Forces Qualification Course, or Special Forces Staff Course (SFSC), or an "S" enlisted MOS identifier.
- Training in the religious background of the region of deployment.

The Nature of Religious Support in Special Forces Units

4-175. The unconventional nature of special operations, the high OPTEMPO, and the multiple mission deployments create stress on units, soldiers, and families. These pressures place special demands for religious support on SF UMTs.

4-176. The USASFC (A) chaplain, under the guidance of the USASOC (A) Chaplain, gives staff supervision to the USASFC (A) religious program. Group chaplains supervise religious support operations and provide technical guidance and resources to subordinate UMTs. SF chaplains frequently deploy to advanced operational bases (AOB). If a Special Forces Operational Base (SFOB) is deployed, a Catholic chaplain, when available, is attached to perform religious support at the SFOB, as well as, at the AOB, and down to the Operational Detachment Alpha (ODA).

Ranger Regiment

4-177. The 75th Ranger Regiment is a major subordinate command (MSC) of the USASOC (A). The Ranger Regiment is capable of conducting both conventional and special operations across the full spectrum of conflict. As a potent, rapidly deployable direct action infantry force, it is particularly well suited to conduct short duration strike operations or other small-scale offensive actions against targets of high strategic or operational importance. Such operations include actions to seize, destroy, or capture enemy forces or facilities, or to recover designated personnel (noncombatant evacuation operations, liberate friendly prisoners of war, capture designated enemy personnel) or equipment from hostile, denied, or politically sensitive, high threat environments requiring exceptionally competent, mature and disciplined forces.

4-178. The Rangers may conduct operations independently, or in support of a campaign plan. While Ranger operations may be conducted in coordination with conventional forces, they differ from conventional operations in degree of risk, operational techniques, and modes of employment. Ranger operations will on occasion be conducted before political preparation of the battlespace has been completed or, in certain cases, before such preparation has even been initiated. Because of its rapid deployment ability and capabilities as a formidable strike force for forced entry operations; the Ranger Regiment maintains an exceptionally high level of operational readiness.

Organization

4-179. The Ranger Regiment's structure is similar to that of a conventional light infantry brigade, but with a robust regimental headquarters that includes organic reconnaissance, signal, military intelligence, and training detachments. Ranger units are not designed for sustained combat and therefore, they are limited by very austere organic logistic support, and by equally limited organic combat multipliers. During all phases of operations and training, Ranger units require responsive and adequate support, either from theater Army assets or USASOC (A).

Unit Ministry Team Personnel Qualifications

4-180. Religious support to Army Rangers is physically, emotionally and spiritually demanding. Unique qualifications include—

- Security clearance for all UMT members equal to mission classification.
- Airborne qualification.
- Ranger qualification.
- Completion of the Ranger Orientation Program (ROP) by chaplains and assistants E-5 and above.
- Completion of the Ranger Indoctrination Program (RIP) by chaplain assistants E-1 through E-4.

The Nature of Religious Support in Ranger Units

4-181. All Ranger elements, from the single platoon to the entire Ranger Regiment, are capable of conducting extremely decentralized training exercises and combat operations. Some additional religious support planning requirements include:

- A thorough understanding of the operational and strategic mission.
- Crisis action planning and execution of religious support

4-182. The Ranger Regimental chaplain exercises staff supervision over regimental religious support and subordinate Ranger battalion UMTs.

U.S. ARMY CIVIL AFFAIRS AND PSYCHOLOGICAL OPERATIONS COMMAND (AIRBORNE)

4-183. U.S. Army Civil Affairs and Psychological Operations Command (Airborne) (USACAPOC (A)) provides civil affairs (CA) and psychological operations (PSYOP) activities and information for both SF and conventional commanders. The CA/PSYOP mission is to inform commanders on the nature of foreign audiences to influence the behavior of foreign governments, organizations, groups, and individuals. To influence a target audience's thinking and behavior, the CA/PSYOP unit must be extremely well versed in the language, culture, thought patterns, and belief systems of the people in the AO.

PSYCHOLOGICAL OPERATIONS UNITS

Organization

4-184. Psychological Operations (PSYOP) personnel often deploy for extended periods as individuals and small teams to support SOF and conventional forces. Deployed personnel are required to work with high-level leaders of other armed forces, and with government officials from the U.S. and other nations.

Psychological Operations Unit Ministry Team Personnel Qualifications

4-185. Requirements for members of UMTs in PSYOPS varies between components to include:

- Security clearance for all UMT members equal to the mission classification.
- Airborne Qualification.

The Nature of Religious Support in Psychological Operations Units

4-186. Tailored PSYOP units in support of SF and conventional forces require flexible religious support. UMTs deploy with a battalion-size or larger PSYOP Task Force (POTF) or PSYOP Task Group (POTG). When POTF or POTGs do not deploy, the religious support is coordinated through the Supported unit.

- PSYOP UMTs frequently coordinate religious support with joint or combined forces.
- When a PSYOP UMT does not deploy, the supported unit provides religious support to PSYOP personnel.
- PSYOP UMTs participate extensively in family support programs during all stages of deployment.

CIVIL AFFAIRS UNITS

4-187. Civil Affairs (CA) units support the commander's relationship with civil authorities and the populace, promotes mission legitimacy, and enhances military effectiveness. CA units support all USASOC (A) missions with regional orientation briefings and information on political, social, cultural, religious, and economic aspects of the Joint Special Operations Area (JSOA).

Organization

4-188. CA units are task organized to support USASOC (A) and supported conventional units' mission requirements. Some CA units are organized as battalions with FID and UW support missions. Other units are organized into special teams, such as Cultural Affairs teams.

Unit Ministry Team Personnel Qualifications

4-189. The assignment of a Unit Ministry Team requires unique qualifications, to include—

- Security clearance for all UMT members equal to the mission classification.
- Airborne qualification.
- Extensive knowledge of religious and cultural issues and their effect on operations.

The Nature of Religious Support in Civil Affairs Units.

4-190. UMTs serving within USACAPOC (A) primary missions are to provide and perform religious support to the unit assigned in accordance with U.S. Code Title X. Civil Affairs UMTs may provide all UMTs in an AO a unique avenue of information and expertise on religious cultures in the AO. CA UMTs can also coordinate for command directed CA activities involving liaison between the command, local religious leaders, and UMTs. CA teams attached to supported commands are provided religious support through the supported command UMT. CA UMTs serve as another religious support asset for the combatant command chaplain's religious support plan.

MEDICAL UNITS

4-191. Deployable Army Medical units are in the process of transformation under the Medical Reengineering Initiative (MRI). MRI is expected to be complete by FY 2009.

4-192. Under MRI, deployable medicine includes C2 units, corps medical augmentation units, and hospitals. MRI C2 units include Medical Brigades, and Medical Commands (MEDCOMs) at Corps and Echelons Above Corps. MRI hospitals come in one type: the Combat Support Hospital, which can be located at Corps and Echelons Above Corps.

Organization

4-193. Unit Ministry Team personnel are organically assigned to MRI C2 units and to MRI hospitals. Corps medical augmentation units receive their religious support through area coverage.

4-194. Since health care skills are highly perishable and in great demand, the largest number of the Army's health care providers are assigned to TDA Medical Treatment Facilities (MTFs). When needed for operational requirements in deployable units, these health care providers fill necessary slots in deploying MTOE medical units under the Army Medical Department (AMEDD) Professional Officer Filler System (PROFIS).

4-195. Likewise, since medical ministry and pastoral care skills are also highly perishable and in great demand, Unit Ministry Team personnel may be assigned to TDA MTFs with PROFIS obligations to deployable MTOE medical units.

Unit Ministry Team Personnel Qualifications

4-196. Qualifications for Unit Ministry Team personnel serving medical units include:

- 7S, Combat Medical Ministry Specialist and
- 7R, Hospital Ministries Clinician for Chaplains;
- Emergency Medical Ministry Course (EMM) for chaplain assistants.

4-197. To meet the complex demands of combat medical ministry and pastoral care, the AMEDD Center and School's Chaplain Training and Operations Branch oversees an institutional training program to prepare Unit Ministry Team personnel providing religious support in medical unit. This program includes—

4-198. Emergency Medical Ministry Course (EMM) is the entry-level course, training chaplains and chaplain assistants in the basics of wartime emergency medical ministry. It is a two-week TDY course taught at AMEDD Center and School, with practical application at Brooke Army Medical Center. EMM is the basic qualifying course for Chaplain Assistants serving in medical units.

4-199. Combat Medical Ministry Course (CMM) is the intermediate course and requires one prerequisite unit of accredited Clinical Pastoral Education (CPE). Also a two-week TDY course taught at AMEDD Center and School, with practical application at Brooke Army Medical Center. CMM focuses upon six core competences: Crisis Ministry; Trauma Ministry; Mass Casualty Ministry; Biomedical and Battlespace Ethics; Grief, Loss, Death and Dying; and Spiritual Hope/Suicide Prevention. CMM is the qualifying course for award of the 7S, Combat Medical Specialist, ASI, the minimum qualification for Chaplains serving medical units.

4-200. Clinical Pastoral Education (CPE): Residency is the advanced level, accredited by the Association for Clinical Pastoral Education, Inc. and administered by the MEDCOM Army CPE System. The Army CPE system offers residencies at Walter Reed, Eisenhower, Madigan and Brooke Army Medical Centers. CPE prepares Chaplains for service within both institutional and deployed medical environments. CPE is required for award of the 7R ASI.

4-201. Clinical Pastoral Education: 7R Trainer (CPE 7R) is the fellowship level. Accredited by CPE, CPE 7R prepares select 7R Chaplains to supervise CPE residency training within the Army CPE System. CPE 7R is required for award of the 56D7R AOC.ASI.

The Nature of Religious Support in Medical Units

4-202. Unit Ministry Teams serving with medical units bring specialized training in spiritual assessment: mass casualty ministry; crisis ministry; grief and loss; spiritual hope; trauma ministry; and ethics to their provision of religious support. They have a dual religious support focus, in caring for the dying and other casualties and honoring the dead, while also nurturing the living in support of the medical staff.

SUMMARY

4-203. The UMT is assigned to a unit at either BN/Squadron or BDE/Regiment/ or Group. The unit is either operationally committed, about to be deployed, returning from deployment, or is in a garrison situation. In garrison, the priorities of religious support must fit in with the command's primary responsibilities:

- Training.
- Family Support.

4-204. Key to performing religious support in this environment is executing staff responsibilities to fulfill various tasks assigned to the UMT in garrison.

4-205. For the operationally committed unit, the UMT is involved in the highest calling for military support: supporting the soldier in combat situations, operations that fall into four categories: Offensive, Defensive, Stability and Support.

4-206. UMTs supporting these units may participate in joint command, and multi-national operations.

RESOURCES

4-207. The following is a list of main resources utilized for writing and updating this doctrine.

- AR 165-1.
- JP 1-05.
- JP 5-00.2.
- JP 3-05.
- FM 16-1.
- FM 3-0 (100-5).
- FM 3-90.3 (71-3).
- FM 4-0.
- FM 25-100.
- FM 25-101.
- FM 71-2.
- FM 100-9.
- FM 100-10-2.
- FM 100-21.
- FM 7-30.
- FM 7-20.
- FM 1-111.
- Lessons Learned from past and current operation
- Lessons Learned from the training centers

Chapter 5

DIVISION UNIT MINISTRY TEAM RELIGIOUS SUPPORT

TACTICAL ECHELON DESCRIPTION

5-1. The division is the largest fixed Army organization that trains and fights as a tactical team. The division echelon possesses great flexibility. The division is a unit of maneuver, organized with varying numbers and types of combat, combat support (CS), and combat service support (CSS) units. It may be armored, mechanized, medium, light infantry, airborne, or air assault. Each are capable of conducting operations over a wide range of environments. Their commanders tailor their organic brigades and attached forces for specific combat missions.

CONTENTS

5-2. The early 1990s brought significant changes to the world and the Army. The end of the Cold War presented threats that were more ambiguous and regionally focused. Force Projection has become our strategy for supporting the power-projection element of U.S. national security. Army divisions are prepared for force projection operations, as part of a joint, interagency, or multinational effort for full spectrum operations.

5-3. U.S. Army corps and divisions conduct Army operations that span the spectrum from peaceful engagement to actual combat. Army divisions train to fight and win battles and engagements as part of a joint force anywhere in the world. Their organization, capability, flexibility, and versatility allow them to conduct operations for stability and support multinational and interagency operations, as well as, in joint operations. The unique capabilities of armored, mechanized, infantry, light infantry, airborne, and air assault divisions provide the U.S. Army the ability to serve the national interests worldwide. (See FM 71-100 for further information.)

AS PART OF A CORPS

5-4. Traditionally, divisions have operated as part of a U.S. corps. In corps operations, divisions normally comprise 9 to 12 maneuver battalions, organic artillery battalions, and supporting CS and CSS units. Divisions perform a wide range of tactical missions and, for limited periods, are self-sustaining.

Corps augments divisions, as the mission requires. A corps or a numbered army may direct a division to perform tasks of operational importance.

AS PART OF A JOINT TASK FORCE

5-5. Division decision-making processes and capabilities make Army divisions effective in joint, multinational, and interagency operations. Army divisions focus on fighting and winning battles and engagements as part of a joint force anywhere in the world.

5-6. The division may deploy as part of a JTF without its traditional corps headquarters and corps support unit. In these types of operations, the division often works directly for the JTF commander. Therefore, divisions must know joint doctrine, tactics, techniques, and procedures. Divisions assigned to a JTF normally conduct traditional tactical operations and may be involved in nontraditional actions, such as interagency operations or operations with the host nation. Divisions are not normally designated as a JTF headquarters.

5-7. A division in a JTF requires experienced liaison personnel. The division may operate with or receive support from joint, multinational, interagency, or NGOs. The division may receive mission support such as joint fires, joint air defense, national level intelligence, and theater and national level communications. Competent liaison to coordinate with these agencies and other headquarters helps ensure properly executed operations.

JOINT TASK FORCE STAFF AUGMENTATION

5-8. Without significant augmentation, a division headquarters has a limited capability to perform as an Army force (ARFOR) or JTF.

DIVISION AS AN ARMY FORCE

5-9. Depending on the mission assigned to the JTF, the Army forces required for the operation may only be a division or a tailored force of units from a division. The division may find that it is the largest land component command group or staff organization assigned to the JTF. In this scenario, the JTF commander may designate a division under its C2 as the Army Forces (ARFOR) headquarters in small-scale contingency (SSC) operations, i.e., the senior Army headquarters within the JTF. Operations include foreign humanitarian assistance, peace operations, noncombatant evacuations, enforcing sanctions, reinforcing key allies, and limited strikes and interventions. Operations are conducted with and in support of government agencies, nongovernmental organizations, regional and international security organizations, and coalition partners. A division designated as an ARFOR executes these responsibilities unless they are specifically excluded. The Army Service Component Command (ASCC), in coordination with the division and DA staff, tailors the forces provided to the ARFOR to meet specific METT-TC conditions anticipated in the JOA. The ARFOR UMT must plan additional RS assets for unit, area, and denominational coverage.

5-10. Designation of the division as the Joint Forces Land Component (JFLC) or the ARFOR of the JTF incurs numerous additional staffing responsibilities. The division will have to coordinate with other JTF-level component or functional commands as well as the service component headquarters of the

JTF establishing authority for administrative and logistic support. The division as ARFOR plans and conducts operations in accordance with the CJTF guidance and detailed plans. The ARFOR will establish liaison personnel to the CJTF, other component commanders, and supporting commanders as necessary or as directed by the CJTF. The division staff will require augmentation to its staff to accomplish the tasks inherent to operating as the JFLC or ARFOR. In the event that a deployed division becomes the JFLC or ARFOR for the JTF, the commander must request staff, CS, and CSS augmentation. The division may need assistance in providing the necessary support units and sustainment activities for prolonged support of significant numbers of assigned or attached forces. (See FM 3-91 (71-100), Appendix F for further information.)

FUNDAMENTALS OF RELIGIOUS SUPPORT OPERATIONS

5-11. The mission of the Division UMT is to coordinate religious support to soldiers, families, and as the force structure evolves, to forward-deployed authorized civilians. Division Chaplains also serve as critical advisors to commanders at all levels of the division providing essential information on soldier and unit morale as affected by religion, quality of life matters, moral/ethical decision making, and the impact of indigenous religion on military operations. The Division Chaplain develops the CMRP and the division RSP on behalf of the Division Commander.

5-12. The battlespace of the 2lst century involves conditions of modularity, split-based operations, increased lethality, non-linearity, greater troop dispersion, increased agility, increased pace of operations, sustainment of tempo and simultaneity of combat operations.

5-13. Combat power in the 21st century will rely heavily upon technology. The cumulative affect of the numerous factors will increase the spiritual, psychological and interpersonal demands placed upon soldiers of the 21st century. Technology can provide additional means to plan, coordinate, and execute religious support to address these demands, but the personal delivery of religious support by the UMT will always be imperative.

ARMY FORCES

5-14. As an Army forces (ARFOR), the impact upon religious support planning and execution is significant. The role of the ARFOR UMT is significantly more diverse and broad than a division operating as a tactical headquarters in a corps. While the division plans, prepares for, and executes many of the same types of activities, its focus is always on the tactical operation in its AO. Unlike the division UMT focus, the ARFOR UMT not only focuses on tactical religious support tasks in its AO, but also provides critical operational-level religious support sustainment to U.S. Forces UMTs and others throughout the JOA. The ARFOR UMT may provide religious support management to tactical and sustainment activities in contiguous or noncontiguous areas, or a combination of both. The ARFOR UMT must focus its planning on multiple, simultaneous actions that accomplish its own, or sustains other's, religious support mission. The tasks performed include—

- Information management including processing, assessment, and dissemination for command and subordinate UMTs from strategic and operational sources.
- Managing religious support tactical missions and tasks to accomplish missions assigned by the Joint Force Commander or Joint Force Land Component Commander.
- Planning and executing self-deployment of religious support assets to ensure comprehensive and seamless religious support.
- Planning and integrating CMO mutual support tasks.
- Coordinating with Special Operations Forces (SOF) UMTs.
- Executing religious support sustainment operations.
- Providing logistical support.

JOINT TASK FORCE

5-15. The joint task force (JTF) chaplain is responsible for planning, coordinating, and implementing all religious support within the JTF for the commander of the JTF. In addition to the already established doctrinal tasks, the JTF UMT is responsible for the following:

- Developing joint religious ministry support plans in support of the operation. This includes preparation of the Chaplain Activities Appendix, under the Personnel Annex of the CJTF's OPLANs, OPORDs, or campaign plans, to include appropriate tabs.
- Identifying RMST requirements such as augmentation, logistics, and communication requirements.
- Coordinating religious ministry with MSC Chaplains.
- Maintain liaison with chaplains of international forces and appropriate host nation (HN) civilian religious leaders.
- Coordinating appropriate training for RMSTs.
- Coordinating with the other staff sections to ensure sufficient religious assets are available to support combat forces, medical services, civilian detainees, EPWs, and others.
- When appropriate and in coordination with the CMOC, the JTF may conduct liaison with faith-based NGOs and IOs.
- Advising distribution of humanitarian assistance supplies from religious organizations.

PLANNING CHECKLIST

- General
  1. Staff: communication, annex, integration, briefings
  2. Specialized:
- Nurture the living:
  - Spiritual Fitness
  - Worship
  - Counseling
  - Visitation to staff and subordinate UMTs
- Care for the Dying:
  - Spiritual triage
  - Leadership
  - Battle Stress
- Honor the Dead:
  - Memorial Services
  - Memorial Ceremonies
  - Rites
  - Sacraments
  - Prayers

Predeployment

5-16. The Division UMT translates the commander's intent and decisions into a fully coordinated and supported religious support operation. Division UMTs, in coordination with their commander and staff, can—

- Accurately determine the current situation.
- Meaningfully assess courses of action for the commander.
- Issue the necessary guidance and instructions, having received guidance from the chief of staff.
- Monitor how well the commander's religious support plan and instructions are being executed.
- Alter the plan in accordance with unit feedback, commander's guidance, commander's intent, and METT-TC.
- Plan future operations.
- Synchronize actions prior to combat operations.

5-17. The division UMT is expected to operate with a great deal of freedom and responsibility. From information available, it must discern what is essential for the commander's decision making. It must coordinate the entire division religious support effort to support the commander's intent and concept of operation. Normally, the commander issues guidance and makes tactical decisions from field locations, then moves to the point of the division's main effort. The UMT develops appendices and plans necessary coordination between subordinate and higher UMTs to execute the commander's intent.

Deployment

5-18. Religious support, like other battlespace support systems, requires C4I and leadership. The Division Chaplain provides UMT C4I and leadership. To perform these duties, the Division Chaplain, like the commander, must be positioned wherever the Division Chaplain needs to be to influence religious support coverage throughout the division area of operations. The Division Chaplain, therefore, cannot be tethered to a specific geographic location but must retain the flexibility to respond to changing conditions as they occur in the battlespace. The Division Chaplain must be able to provide C4I while at the same time provide direct religious support to the commander and staff.

5-19. The operational tempo of conflicts will demand 24-hour operations by the units engaged, religious support operations cannot be an exception to this demand if timely religious support is to be provided. Twenty-four hour operations allow for uninterrupted tracking of the battle. The purpose for tracking the battle is to monitor the fight so as to enable timely responses to emerging battlespace conditions requiring religious support.

5-20. In a division fight, 24-hour operations can be confined to a battle-tracking cell if that cell has the common relevant picture and can communicate to outlying UMTs. These outlying UMTs can then maintain 24-hour availability. Battlespace conditions will require 24-hour operations to monitor both current operations and future operational planning.

5-21. Religious support relies upon the availability of sufficient UMT assets to provide coverage. The capability to man 24-hour battle tracking operations cells and the flexibility to reallocate religious support personnel is dependent upon sufficient religious support resources. Flexibility must include the ability to tailor UMTs to match the force packages engaged in the operations.

Chaplain Detachments-Division

5-22. The Chaplain Detachments-Division (CD-D) allows the Division UMT to conduct 24-hour operations. The Chaplain Detachment will come from the Reserve Component and be organic to the Division it supports. This team consists of one chaplain (MAJ) and one chaplain assistant (SSG). The team will drill, train and deploy with the supported division. (See Appendix G and 6-69) Division Chaplains must incorporate the use of CD-Ds into their RSPs and request their deployment through proper staff actions.

5-23. Flexibility must include provisions to transition offense, defense, stability, and support missions.

5-24. Additional planning considerations include—

- The HQ may be structured for incremental deployment.
- Each deployment increment must be synchronized to ensure comprehensive religious support.
- TPFDD information on augmented religious support assets.
- Religious support assets and resource monitoring and management.
- Rules Of Engagement.

Pre-combat

5-25. This phase for the division UMT requires finalized preparation for combat operations. The UMT works closely with the staff, especially the G3/J3, monitoring the tactical situation for last minute changes to the religious support plan. The chaplain assistant NCOIC ensures that—

- Logistical requirements are finalized.
- Battle tracking requirements for the chaplain section are finalized and prepared.
- Appropriate support to Information Operations is coordinated.
- The NCO support channel is advised and monitored for support.
- Personnel issues for the division chaplain assistants are resolved and a contingency replacement plan is staffed with G-1/J-1.

5-26. UMT planning considerations include—

- Spiritual care to subordinate UMTs.
- Spiritual care to division command and staff.
- Managing the implementation of the religious support plan.
- Monitor the current operations situation applying METT-TC.
- Ensuring support for combat operations religious support plan.
- Division chaplain section personnel are spiritually, emotionally, and physically prepared for the operation.

Operations (See Figure 5-1.)

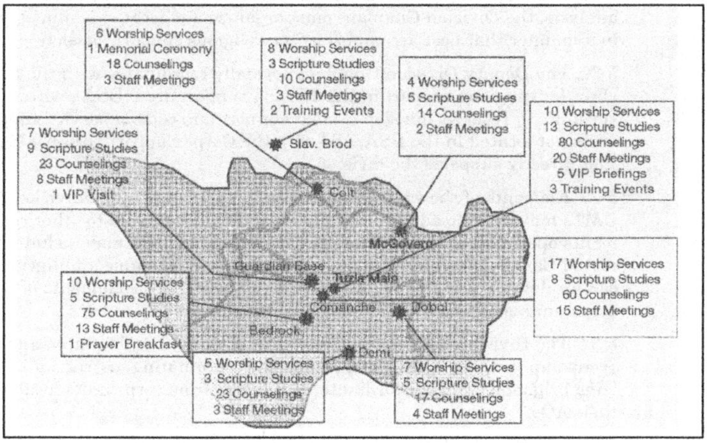

Figure 5-1. Week of Religious Support in Task Force Eagle

5-27. *Critical capabilities requirements.* The ability to provide religious support in the 21st century battlespace will depend upon five critical capabilities requirements: information access, situational awareness, communication connectivity, mobility, and force structure. A deficit in any of these five requirements will radically impair the ability to provide timely religious support.

5-28. Division UMTs must have real time operational awareness. They must be able to determine an appropriate response and expeditiously communicate that information to religious support personnel who will respond directly to the situation. To enable religious support throughout the battlespace, UMTs must be able to move and communicate seamlessly throughout the AO.

5-29. Battle tracking operations require sufficient personnel and the required technical training to conduct 24-hour operations.

Offense

5-30. The Division Chaplain coordinates and supervises religious support for the comprehensive support of all soldiers and units in the division. Based on the religious support estimate, the division chaplain determines area support requirements and where to position the division UMT to best accomplish its mission.

5-31. Normally, the Division Chaplain monitors the tactical situation, plans religious support, and provides care for soldiers from the main Command Post (CP). Prior to an operation, the Division Chaplain moves forward and provides pastoral care to brigade UMTs. Division UMT personnel, under the

Deputy Division Chaplain, provide religious support to soldiers in the rear CP; monitor the volume of casualties, replacements, logistics; and coordinate religious support in the division rear area. These are norms, but in the final analysis, the Division Chaplain must organize and locate his ministry team in a manner that best accomplishes the religious support mission.

5-32. The Deputy Division Chaplain normally coordinates with the DISCOM UMT for religious support in the division support area (DSA); with maneuver units in reserve, division combat support and combat service support units not located in the DSA; and with the Corps Support Group (CSG) UMT that directly supports the division.

5-33. Elements of the supporting CSG may be located in the division AO. Its UMTs may provide additional support to the division and to other corps elements operating in the division area. Other corps units may include the Armored Cavalry Regiment (ACR), artillery, signal, military intelligence, medical, air defense artillery, aviation, and engineers. Some elements of these corps units may be forward in brigade support areas.

5-34. The Division UMT coordinates with the CSG UMT for area and faith group support. The division RSA includes coordinating instructions authorizing brigade UMTs to coordinate with UMTs from corps units located in their AOs.

5-35. The Division UMT coordinates with the Corps UMT concerning area support of corps units in the division AO. The tactical situation may prevent corps UMTs from moving forward to their subordinate units. The teams may be separated from their forward elements by as much as 150 kilometers.

5-36. Area support is limited in light infantry divisions because of restricted transportation assets. Non-divisional UMTs with greater mobility may be tasked to provide more area support to a light division than they would to a heavy division.

Defense

5-37. Division religious support during the defense is similar to religious support during the offense. Access to subordinate UMTs is easier, but enemy reconnaissance and preparatory fires may complicate travel. As in the offense, the division UMT coordinates with the CSG UMT for mutual support.

5-38. The DISCOM UMT coordinates area support in the DSA.

Stability

5-39. Conducting religious support during stability operations remains fundamentally the same as in the offense and defense. The management of the religious support plan is challenged by the increased information management tasks required to properly respond to religious support issues. In stability operations, as in war, the division conducts simultaneous operations. For example, the division could conduct peacekeeping operations at the same time it is deploying additional forces and securing facilities. Stability operations are usually complex. Rarely will UMTs be given a single task to accomplish. Their capability to handle multiple tasks across a wide range will challenge them. When the division headquarters is deployed for stability operations, it is often assigned as the ARFOR and works for a joint task force commander.

5-40. Although often only part of the division deploys, as a controlling headquarters, it may receive attached or operational control (OPCON) units from many external sources. These sources include Army organizations, other U.S. military services, and multinational forces. Within these units are other UMTs that the division UMT must integrate into their religious support plan. Planning considerations for the division UMT—

- ROE understanding and religious support application.
- Command understanding of the religious culture and traditions.
- Media and Public Affairs Office/Officer (PAO) guidance.
- Religious support integration during Base Camp planning.
- Command and staff expansion for operations.
- Information management.
- Personnel issues for UMTs
- Increased liaison tasks with faith-based NGOs and IOs, and host nation religious leaders.

Support

5-41. Support operations consist of relief operations, support to WMD incidents, support to civil law enforcement, and community support. All are challenging for religious support planning and execution. Support operations can also be conducted simultaneously with offense, defense, and stability operations. Posse Comitatus (the legal guidelines defining the relationship between federal armed forces and the civilian populace) plays a critical part for planning. (See Annex K for more information.)

5-42. Religious support planning may include—

- Limits on deployed religious support personnel due to statutory agreements and other restrictions.
- Availability and type of transportation assets.
- Force protection requirements.
- Training requirements.
- SOF participation.
- Use of reserve component forces.
- Other U.S. government agencies' roles.
- Lodgment sites and responsibilities.
- Host nation requirements and expectations (OCONUS).
- Political and social implications of the division's actions.
- Funding and special equipment and supplies requirements.
- Impact of mass casualty upon the units.

Reconstitution, Redeployment, and Demobilization

5-43. Reconstitution is an extraordinary action that commanders plan and implement to restore units to a desired level of combat readiness. It transcends normal daily force sustainment actions. No resources exist solely to perform reconstitution; rather existing systems and units do it.

5-44. Reconstitution decisions belong to commanders. They control assets and decide whether to reorganize or regenerate a unit. Normally, the decision to reconstitute is made two echelons above the unit that must be rebuilt. The commander two echelons above, with his staff's assistance, are in the best position to assess a subordinate unit's effectiveness. The division UMT provides unit ministry support, particularly for cases of combat and operational stress, spiritual renewal, critical event debriefings, and worship opportunities. The chaplain coordinates worship and memorial services, sacramental acts, and pastoral counseling. Chaplains play a key role in assessing the spiritual morale and moral/ethical needs of the unit.

5-45. The OPLAN includes enough details to enable staffs and supporting units to prepare for rapid restoration of units within command priorities. The division UMT inputs the religious support plan in support of reconstitution operations after utilizing the estimation and analysis process. While the religious support plan cannot meet all the contingencies of Army operations, it must be one that the command and UMT can adapt to the situation. In addition to addressing reconstitution for a specific mission, the staff may have to write a separate OPLAN for reconstitution operations. The UMT is fully integrated within this process. The more fully developed a UMT's reconstitution SOP, the easier for it to develop a reconstitution plan.

5-46. Redeployment is the repositioning of forces after mission accomplishment or transfer of mission and authority to another unit. This may mean redeployment to another AO for subsequent missions, or return to home station. In either case, the division UMT should prepare subordinate UMTs to assist commanders, soldiers, and family members, in preparing emotionally and spiritually for the changes that come with redeployment and reunion.

5-47. Demobilization is the return of forces to a peacetime posture. Planning for demobilization is critical. The division UMT cannot wait for the operation to end to begin planning for demobilization. It must begin early in the process. The division UMT's role in demobilization may be much greater than in mobilization. It begins in the division's AO with reconstitution of attached Reserve Component forces and preparation for redeployment to CONUS or OCONUS demobilization stations. Demobilization planning begins concurrently with mobilization. Demobilization planning may impact on division UMTs in theater as reserve component units begin the tasks of reconstitution and demobilization, while under division or corps control. Religious support plans must address many issues including—

- UMT spiritual reconstitution.
- Decorations and awards.
- Evaluation reports.
- Memorial Services and Ceremonies.
- Critical event debriefings.
- Reunion briefings for command and staff, soldiers, and family members.
- The division UMT may also coordinate with the corps and installation chaplain at both the division's home station installation and the RC unit's demobilization station.

Garrison Operations

5-48. During garrison/installation operations, the division UMT plans and implements the CMRP in coordination with the unit training schedule to support the religious needs of the division's soldiers and their families. The division UMT manages and supervises the subordinate UMTs assigned to the division for the commander and coordinates ministry as applicable with the garrison/installation UMT. The UMT participates in all installation chaplains' training programs and other programs that enhance the religious quality of life of the installation. They also fully participate in the unit's training program. The division UMT will support the unit's training program by being available to conduct training for the command and staff, and subordinate UMTs. The division UMT is the mentor for the battalion/squadron UMTs of the division and will execute a mentorship program. Also, the senior division UMT supervises the personnel assigned or attached to the division chaplain's office.

ROLE OF THE DIVISION UNIT MINISTRY TEAM

5-49. The role of the division UMT is critical to the professional and tactical proficiency of the subordinate UMTs. Their leadership responsibilities shape and guide brigade, battalion, and squadron UMTs role in religious support execution process. They are the command and staff advisors for the officers and NCO chain of command.

PRINCIPLES

Table 5-1. Division/Task Force Tactical UMT

CHAPLAIN (O-5) ***CHAPLAIN (O-4)***	***CHAPLAIN ASSISTANT (E-8)*** ***CHAPLAIN ASSISTANT (E-5)*** ***CHAPLAIN ASSISTANT (E-3)***
Role Commissioned Officer/Staff Religious Professional Title 10 Protected	***Role*** Noncommissioned Officer Specialized Assistant Religious Support Specialist Team Security
Mission Advises commander on religious support and quality of life issues to support the unit's mission. Performs/provides/coordinates the religious support mission for the division. Division Team Leader for Religious Support.	***Mission*** Advises Chaplain and NCO Chain. Provides specialized assistance in areas of religious support to support the chaplain. Perform Battle Staff functions. Team NCOIC.
Parallel Tasks 1. Assess subordinate UMTs readiness. 2. Conducts staff assistance visits. 3. Determines Training Objectives. 4. Mentors subordinate UMTs. 5. Recommend personnel assignments.	***Parallel Tasks*** 1. Assess subordinate UMTs readiness. 2. Conducts staff assistance visits. 3. Conducts UMT religious support training. 4. Mentors subordinate UMTs. 5. Recommends Chaplain Assistant assignments.

5-50. The Division Chaplain is a personal staff officer, religious leader, and team leader. The Division Chaplain Assistant NCOIC must be a NCO leader and advisor, mentor, information manager for the team. They both must have a keen awareness of the world situation as it bears on potential military operations, know the role of military religious leader's and have a broad understanding of the impact of indigenous religions on military operations.

5-51. Unit Ministry Teams must use technology to enhance the personal delivery of religious support. They must be trained in joint and multi-national operations and skilled in synchronizing all aspects of religious support during military operations. They will operate in an environment of increasing volatility, complexity and ambiguity. The division UMT must be innovative and disciplined.

5-52. This means that the Division Chaplain section NCOIC will need to be a graduate of the Battle Staff NCO Course (with the 2S identifier). This will enable the UMT NCOIC to take part in the MDMP in the absence of the chaplain, track the battle analytically and better utilize the digital enablers that the division will share with other staff sections.

5-53. Division UMTs require the same level of technology as the command of the supported force. UMT automation, communication and transportation requirements must keep pace with the command's requirements in order to effectively coordinate religious support, communicate critical information, and remain physically flexible to meet requirements across the division battlespace. They are an integral part of the division.

RESPONSIBILITY AND ACCOUNTABILITY

Staff Responsibilities

5-54. The Division UMT works directly for the commander representing the religious support dimension of caring for soldiers and family members. The UMT executes the commander's plan for religious support initiatives and programs. Staff responsibilities include—

- Synchronizing religious support programs across the division AO.
- Coordinating denominational coverage.
- Performing UMT personnel tasks.
- Advising the Division Commander on matters of religion, ethics, morals, and morale as affected by religion, the impact of indigenous religions on military operations, and installation and community religious support issues.
- Preparing the staff religious support estimate and appendix for the division OPORDs.

Relationship to the Corps Unit Ministry Team

5-55. The Division UMT supports all of the Corps UMT programs including training, family support, and professional development. They work closely together implementing UMT personnel issues for the command. During combat operations, the division UMT utilizes the Corps UMT's religious support appendix to the corps OPORD for religious support planning. The Division UMT works closely with the corps to ensure the success of the religious support mission.

SUPERVISORY RESPONSIBILITIES

5-56. The Division UMT provides technical and professional supervision to subordinate UMTs in the division. As the senior UMT, the Division UMT provides leadership for UMTs in professional development, implementation of the CMRP, personnel issues, and other issues as directed by the commander. The Division UMT works closely with subordinate UMTs to mentor and provide direction.

TRAINING

5-57. The Division UMT assesses its current training strategies as well as the underlying assumptions regarding pre-training knowledge and skills in light of the Division Commander's training and mission requirements. The Division UMT will provide training in planning, preparation and execution of the 11 Religious Support Activities in accordance with the CMRP. Information literacy requirements for all UMTs are imperative in training.

5-58. Training will be profoundly affected since non-linear operations complicate the organization of the battlespace into close, deep and rear operations.

SUMMARY

5-59. Recent significant changes in international politics have created an ambiguous security environment in which threats to world stability and U.S. national interests are more regionally focused. Many crisis situations in which the United States exercises its power span the spectrum of operations from operations other than war to actual combat. Force Projection has become our chief strategy for supporting the power-projection element of U.S. national security. Army corps and divisions must be prepared for force projection operations, as part of a joint or multinational effort and in war as well as other than war environments.

5-60. The mission of the Division UMT is to coordinate religious support to soldiers, families, and as the force structure evolves, to forward-deployed DOD authorized civilians. Division Chaplains also serve as critical advisors to commanders at all levels of the division providing essential information on troop and unit morale, quality of life matters, and the impact of indigenous religion on military operations.

5-61. The Division UMT is expected to operate with a great deal of freedom and responsibility. From information available, it must discern what is essential for the commander's decision making. It must coordinate the entire division religious support effort to support the commander's intent and concept of operation. The UMT develops plans and conducts necessary coordination between subordinate and higher UMTs to execute the commander's intent.

5-62. Combat power in the 21st century will rely heavily upon technology. The cumulative affect of the numerous factors will increase the spiritual, psychological and interpersonal demands placed upon soldiers of the 21st century. Technology will provide additional means to plan, coordinate, and execute religious support to address these demands but the personal delivery of religious support by the UMT will always be imperative.

RESOURCES

5-63. The following is a list of main resources utilized for writing and updating this doctrine.

- JP 1-05.

- JP 5-00.2.
- JP 3-05.
- FM 16-1.
- FM 100-5.
- FM 3-0 (100-5).
- FM 25-100.
- FM 25-101.
- FM 71-100.
- Lessons Learned from past and current operation
- Lessons Learned from the training centers

Chapter 6

CORPS/ECHELONS ABOVE CORPS RELIGIOUS SUPPORT OPERATIONS

The corps is the Army's largest tactical organization. The corps organization is tailored for the assigned theater of operations and the assigned mission. Corps may be assigned divisions of any type required by the theater and the mission. They have a support command and will receive additional Combat Service (CS) and Combat Service Support (CSS) organizations based on the assigned mission.

Echelons above corps (EAC) consist of the Army at theater level and above. These units will be under the command of a regional combatant commander (unified command). This level is in the strategic context.

CONTENTS

OPERATIONAL ECHELON DESCRIPTION

6-1. Corps conducts operations with joint, interagency, or multinational forces, facilitated by information superiority across the full range of military operations, in a Major Theater War or small-scale contingency, against any threat, in any assigned AO. Corps operations may include a mixture of offense, defense, stability, and support operations executed sequentially or simultaneously.

6-2. Corps is usually the link between the operational and tactical levels of war. Corps may link the operational level of war to the strategic level of war. Corps plan and conduct major operations and battles. The corps headquarters synchronizes tactical operations including maneuver; fires, joint, deep, close, and Suppression of Enemy Air Defenses (SEAD); CS; and CSS. The corps is not a fixed standing organization, but a formation whose core characteristic is the ability to form task forces focused on mission requirements. It employs tailored forces that are optimized for specific functions or tasks. The corps employs divisions and applies force allocation rules of support to divisions based on mission. The corps normally has a wide range of Army capabilities to support and sustain any effort. The two critical requirements of corps operations are a robust C4 intelligence, surveillance and reconnaissance (C4ISR) and logistics support to plan, prepare, execute, and assess the full range of operations. Corps traditionally can employ multinational forces and operate with governmental and non-governmental agencies and capabilities. Corps operations are nested in political military dimensions of the combatant commander's theater effort, but are optimized for conducting military operations.

6-3. The corps is a much diversified organization that fulfills the synchronization of many functions that achieve tactical and operational advantage. The corps conducts planning together with joint forces to achieve campaign objectives and conducts simultaneous operations in depth across the corps battlespace. The corps integrates all joint combat, CS, and CSS efforts.

6-4. The corps plans and conducts campaigns and major operations. Corps campaign plans set broad long-term goals, such as control of a geographical area, defeat of an enemy in theater, or the reestablishment of international boundaries. The corps also plans and executes tactical-level battles that support the corps operational objectives, simultaneously conducting close, deep, and rear operations, as well as planning future operations.

6-5. Current U.S. collective security arrangements make it imperative that corps are ready to deploy in response to worldwide situations. In the event of a crisis situation, a corps headquarters considers the employment of light forces, armored maneuver forces, and special operations forces. (See Figure 6-1.)

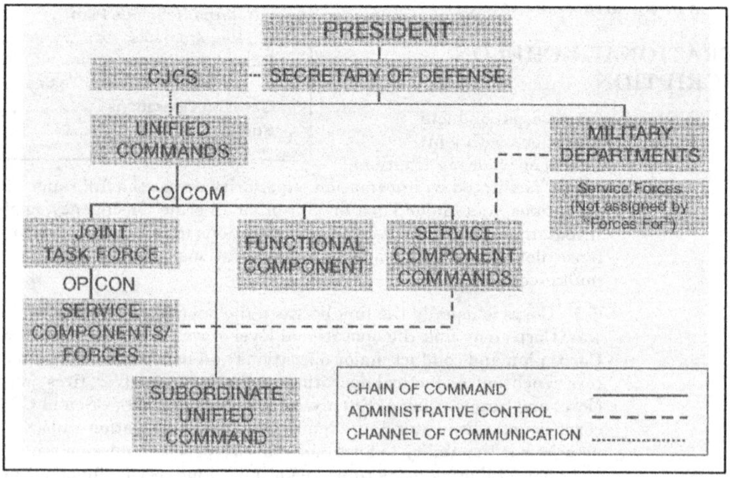

Figure 6-1. Chain of Command and Control

JOINT TASK FORCE CONFIGURATIONS

6-6. A corps headquarters may function as the Army component headquarters of a joint force, the Joint Forces Land Component Command (JFLCC) headquarters of a JTF or as the JTF headquarters itself. The corps' primary responsibility in the joint environment is to participate in the development of supporting joint campaign plans. Current U.S. collective security commitments and the very nature of the international political environment ensure that U.S. corps' will also operate as part of multinational coalition forces.

6-7. Corps with missions to respond to worldwide situations will frequently fight as the largest ground formation in the theater. Often this situation results in the corps commander filling the position of the commander JTF (CJTF). In such cases, the corps headquarters accepts the responsibility to plan operational and tactical level missions as well as oversee the tactical execution of campaigns or other extended operations.

6-8. As the CJTF, the corps commander will exercise OPCON or tactical control (TACON) of assigned forces. The CJTF must form a joint staff and establish the necessary joint boards and centers, as the mission requires. Joint service representation on the staff should be in relative proportion to the service composition of the JTF. It is imperative that all of the staff sections and agencies have joint service representation in the formation and execution of the joint campaign plan. The corps staff must transition to a joint staff structure to ensure unity of effort within the task force.

6-9. The JTF establishing authority is responsible for providing personnel and resources for the corps when the corps is designated the JTF headquarters. However, the corps commander, as the CJTF, must determine what augmentation requirements he needs for the task at hand and coordinate support through the establishing authority. This augmentation is essential in the transition of the corps to a JTF structure.

6-10. Augmentation must be tailored for the specific situation and will be different for every mission. The following areas usually require augmentation:

- Joint, special staff, and personal sections.
- Specific functional area augmentation, such as CA, movement control, and PSYOP.
- Headquarters life support functions.
- Communications support.
- Security support for the JTF headquarters

6-11. The combatant commanders usually organize augmentation for the joint staff into packages based on the nature and tempo of developments and the existing capabilities of the headquarters designated as the JTF. These packages fall into three broad categories:

- Quick response cells to rapidly augment the JTF headquarters with joint and operational area expertise during the early stages of a crisis action situation.
- Functional area augmentation to provide the JTF with expertise not organic to its staff or to enhance specific capabilities based on mission requirements

- Joint staff augmentation to provide the coordinating personal and special staff sections expertise in joint procedures and service-unique capabilities.

JOINT TASK FORCE STAFF AUGMENTATION

6-12. Augmentation facilitates coordination between the JTF staff and the combatant command joint staff, especially with regard to accessing information and capabilities available at the combatant command level.

6-13. Although the corps/ JTF commander applies the military element of power, he also considers those interagency, political, economic, and other resources that may be available and appropriate for the task at hand. The CJTF must understand the strategic and regional environment to include U.S. policy, treaty commitments, the status of forces agreements (SOFA); host nation concerns, religious and cultural considerations, and coalition members' interests. These influences will affect campaign and operational planning, and the establishment of ROE for the joint force. The CJTF must keep in mind the strategic end state through out the planning and execution of the operation.

THE CORPS AS AN ARMY FORCES/JOINT FORCE LAND COMPONENT COMMANDER HEADQUARTERS

6-14. A JTF-level service component command consists generally of forces from only one service of the JTF. The CJTF may designate the corps headquarters as the ARFOR headquarters giving it Title 10 responsibilities and, in most cases, OPCON of a number of assigned or attached forces.

6-15. The designation of Commander, ARFOR carries additional responsibilities for the Corps Commander. The corps as ARFOR must accomplish the same coordination as covered above. Corps headquarters are uniquely suited to be the ARFOR or Joint Force Land Component Commander (JFLCC) headquarters in a JTF. Joint Pub 1-02 defines the Joint Force Land Component Commander as "the commander within a unified command, subordinate unified command, or a JTF responsible to the establishing commander for making recommendations on the proper employment of land forces, planning and coordinating land operations, or accomplishing such operational missions as may be assigned. The joint force land component commander is given the authority necessary to accomplish missions and tasks assigned by the establishing commander. The joint force land component commander will normally be the commander with the preponderance of land forces and the requisite C2 capabilities." The primary purpose of a JFLCC is to provide unity of command for employing land power for the benefit of the joint force as a whole and to give the Joint Force Commander (JFC) a workable span of control.

6-16. The JFLCC must shape the battlespace for subordinate units by focusing on deep operations in space and time (more than 96 hours out).

6-17. Subordinate units are organized for executing the battle command functions required from the higher headquarters of ground maneuver forces. Corps headquarters can coordinate the sustainment activities for significant numbers of assigned or attached forces. Corps also has the staffing to do joint and multinational force coordination and liaison. In most cases the corps headquarters is capable of being the ARFOR or JFLCC headquarters in a JTF without significant personnel augmentation.

THE CORPS AS A FORCE PROVIDER

6-18. In day-to-day operations the corps is required to provide resources to support operations in which the corps headquarters is not otherwise involved. (As an example, each U.S. corps today has soldiers deployed around the globe.) As a force-provider, corps has considerable assets, which they may commit to other commands to support major and lesser regional contingencies around the world. Although these deployed organizations may be under the OPCON or TACON of another headquarters, the corps retains significant responsibilities as the parent organization.

6-19. Operations in stability and support environments are good examples in which corps' function as force providers. Stability operations and support operations often come with vague or ad hoc chains of command, requiring specific support. These missions are different. They require a variety of corps combat, CS, and CSS units to operate in concert with U.S. and foreign civil agencies of government, international, and private organizations.

ECHELONS ABOVE CORPS

6-20. EAC echelons include the theater command. Within this command are various commands that support theater operations. The Army Service Component Command (ASCC) is an Army component in a unified and subordinate unified command. Army forces, through the ASCCs, provide defenses to all joint force components. The ASCC establishes the link between ARFOR and the joint command. It also plans and executes operations in support of the joint campaign, plans and executes support operations to sustain subordinate ARFOR, and provides support to other services in accordance with executive agent responsibilities. The Theater Support Command (TSC) is an EAC-level support organization that is key to operations for CS and CSS execution. The TSC is structured and organized to deploy tailored functional modules. The ARFOR commander is provided key functional expertise and C2 to support joint forces.

FUNDAMENTALS OF RELIGIOUS SUPPORT OPERATIONS

6-21. The corps can rapidly tailor force packages formed from the full range of Army capabilities, both active and reserve. The corps C2 has the organic capabilities for split-based operations required for simultaneous functions of deploying and employing Army forces. The religious support planning and management mission considers force-tailoring UMTs to perform and provide religious support for all types and sizes of forces in all contingencies. Religious support delivery, coordination, synchronization, and synergy are dependent upon the communication capacity in both digital and analog environments to support decisive operations and to improve the synergy of joint and multinational operations. Superior situational understanding is available through information systems that facilitate the execution of decisive, shaping, and sustainment operations across all dimensions of the corps AO.

CORPS RELIGIOUS SUPPORT

6-22. The Corps Chaplain supports the Corps Commander's operational-level responsibilities and roles by engaging in three operational areas:

- Given guidance by the Corps Commander in coordination with G-5, PAO, CA, the Corps Chaplain establishes links with representatives of joint, multinational, interagency, faith-based NGOs and IOs, and religious leaders of the host nation.
- Plans and executes religious support for corps operations. Monitors religious support in major subordinate commands.
- Executes support operations to sustain subordinate Army forces. (See FM 100-7.)

6-23. From the main CP, the Corps UMT plans and synchronizes religious support in the corps area. The Corps Chaplain monitors the tactical situation and moves as far forward as possible to provide staff supervision. The Corps UMT at the rear CP recommends chaplain and chaplain assistant assignments. It coordinates religious support for displaced persons and also plans religious support for reconstitution. It provides an on-call UMT for the Corps Reconstitution Task Force (RTF).

6-24. The Corps UMT normally passes responsibility for religious support in the corps support areas (CSA) to the COSCOM UMT. The Corps UMT must be capable of simultaneous employment and deployment. Its flexible METT-TC planning speeds the process of force tailoring, while its capabilities allow it to provide religious support to the reception, staging, onward movement, and integration (RSOI) at the lodgment base or in an intermediate staging base (ISB) within a JOA. The Corps Chaplain's section will have secure, robust, and survivable communications networks for transmissions of voice, data, and video.

6-25. The senior echelons above division (EAD) Chaplain will tailor the staffing of religious support personnel to match force packages engaged in operations. Religious support relies upon the availability of sufficient religious support assets and the flexibility to reallocate those assets.

6-26. Flexibility must include provisions to transition between combat and other military operations. Planning considerations include—

- Provide interface/liaison between corps and local religious leaders and faith-based NGO/IOs.
- Support the resettlement of refugees and displaced civilians.
- Reconstitute and regenerate combat ineffective units.
- Manage religious support personnel assignments.
- Contractor religious support planning for subordinate UMTs.
- In coordination with the ASCC chaplain, establishes liaison with senior UMTs of other services for religious support to Army authorized personnel, as well as, religious support to service members of other services.
- Conduct specialized religious support assessment of the AO for the commander and staff.
- Coordinate all chaplains' fund activities within the corps and all budget and funding requirements for the corps UMTs.
- Provide spiritual care to subordinate UMTs.

- Coordinate with the ASCC chaplain for Chaplain Detachments (CD) and Special Medical Augmentation Response Team-Pastoral Care (SMART-PC) usage.
- Provide a UMT training program.
- Plan RS 96 hours to 5 days from execution for the corps fight.
- Utilization of "requests for information" (RFI) to respond to FRAGOs and mission RS implied tasks.

6-27. The Corps Chaplain's staff will be resourced to conduct continuous operations. The Corps Chaplains staff will consist of at least four (4) chaplains and five (5) chaplain assistants and will be augmented by the Chaplain Detachment – Corps. Further augmentation may be necessary if serving as the JTF or JFLCC UMT. The Corps UMT will execute sleep plans to minimize the degrading effects of sleep loss. The Chaplain's section must be equipped to minimize the stress imposed by extreme environmental conditions. The Corps UMT will have the capability to exercise religious support management for the commander from any point within the corps extended area of operations.

THE CORPS/EAC TACTICAL/OPERATIONAL/STRATEGIC UMT	
CHAPLAIN (O-6)	***CHAPLAIN ASSISTANT (E-7/8/9)***
Role Commissioned Officer / Staff member Religious Professional Title 10 Protected	***Role*** Noncommissioned Officer Specialized Assistant Religious Support Specialist Team Security
Mission Advises commander on religious support and quality of life issues to support the unit's mission. Performs/provides/coordinates the religious support mission for the command. Senior Team Leader	***Mission*** Advises Chaplain and NCO Chain. Provides specialized assistance in areas of religious support to support the chaplain. Perform Battle Staff functions. Senior Team NCOIC
Parallel Tasks 1. Conducts staff assistance visits. 2. Determines Training Objectives. 3. Mentors subordinate UMTs. 4. Recommends personnel assignments.	***Parallel Tasks*** 1. Conducts staff assistance visits. 2. Conducts religious support training. 3. Mentors subordinate UMTs. 4. Recommends Chaplain Assistant assignments.

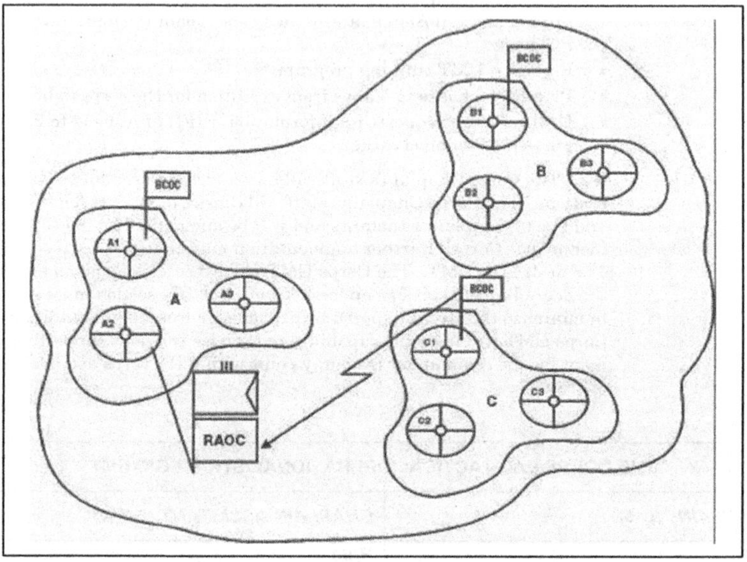

Figure 6-3. Religious Support Coordination in a Corps Support Area

CORPS SUPPORT COMMAND

6-28. The Corps Support Command (COSCOM) UMT coordinates religious support in the corps support area (CSA) for the Corps chaplain. The CSA constitutes the Corps Rear Area. The potential expansion of the CSA, the dispersion of units in bases and base clusters, and the movement of small units make the coordination of religious support difficult. The COSCOM UMT will manage and coordinate area religious support for the Base Cluster Commander to ensure synchronized RS to all units in the base footprint. As a measure of the magnitude of possible expansion, during Operation DESERT SHIELD/DESERT STORM, one COSCOM grew to over 60 UMTs. The area that the COSCOM may cover encompasses the entire battlespace, making authorized and assigned transportation and communication assets for all COSCOM UMTs essential.

6-29. The COSCOM UMT works closely with the Rear Tactical Operation Center (RTOC) and Rear Area Operation Center (RAOC) to track the locations and faith groups of the chaplains in the CSA.

6-30. The COSCOM UMT monitors the flow of religious support supplies with the Corps Material Management Center. The COSCOM UMT locates where communication, battle tracking, and religious support management are best executed. Twenty-four-hour operations are critical to ensure comprehensive religious support and personal staff support to the commander. Split operations at two locations may be considered.

CORPS SUPPORT GROUP

6-31. The COSCOM supports the corps with two or more Corps Support Groups (CSGs): forward and rear. CSGs are composed of multi-functional battalions that provide supply, services, and maintenance to supported units. Forward CSGs support the divisions and other corps units in their AOs. The Rear CSG supports the corps rear area, units in reserve, separate brigades, and units with special missions. The task organization of a CSG can change daily.

6-32. The Forward CSG UMTs monitor the rapidly changing situation through their assigned RAOCs. The tactical communication system found in the RAOC is the primary means of communication for the CSG UMTs. These teams are responsible for coordinating religious support throughout their AOs. They maintain a database with the locations of all UMTs by unit and faith group to process requests for area and faith group religious support.

6-33. The Rear CSG UMT provides religious support to the largest unit in the corps support area that includes the corps and COSCOM headquarters.

6-34. The Rear CSG UMT is often given the reconstitution mission.

RELIGIOUS SUPPORT IN REAR AREAS

6-35. The dispersion of units in rear areas presents a significant religious support challenge. Unit ministry teams have elements of units deployed far forward and out of reach. At the same time, they have elements of other units located in their areas. Of necessity, area support in the rear consumes a greater portion of a UMT's time and energy than does unit support. The Corps Support Chaplain has the responsibility for the rear. Timely religious support coordination is crucial and requires a clear understanding of rear operations.

REAR OPERATIONS

6-36. Rear operations support tactical forces in contact, guaranteeing freedom of maneuver, continuity of support, and uninterrupted C2. In the linear battlespace these actions occur behind forces engaged in active combat. In the non-linear battlespace they occur in the support base areas. Rear operations are characterized by a wide dispersion of multi-functional units conducting a variety of missions around-the-clock.

6-37. Rear areas may be small in a sparse theater, as in small-scale contingencies (SSC), or large in a mature theater, as during general war. In SSC, with tactical operations arrayed through 360 degrees, the "rear area" refers to the area designated as the support base.

6-38. The synchronization of rear operations is the responsibility of a Rear Tactical Operations Center (RTOC), a Rear Area Operations Centers (RAOC), or a Rear Command Post (RCP).

6-39. The Rear Tactical Operations Center (RTOC) is the primary staff element for planning and coordinating rear area security operations.

6-40. The Rear Area Operations Center (RAOC) is a subordinate command post in or near the rear tactical command post concerned with terrain management and security operations.

6-41. The RAOC maintains critical information for the UMT to accomplish its mission: a current situation map; continuous communications with the RCP, other RAOCs, and subordinate base clusters; the threat condition; and Base Defense Status Reports.

> The RAOC maintains critical information for the UMT.

6-42. The most important RAOC function for religious support coordination is terrain management. The RAOC manages the use of terrain by assigning units to bases and base clusters.

Bases

6-43. Units are positioned in bases to share security responsibilities and capitalize on mutual strengths. Bases have clearly defined, defensible perimeters and established access controls.

6-44. Normally, the senior commander is the Base Commander. All forces assigned to the base are under his operational control for defense purposes. The commander forms a Base Defense Operations Center (BDOC) to assist in planning, coordinating, integrating, and controlling base defense efforts.

6-45. The base commander's UMT obtains information about the units (including those in transition) occupying the base from the BDOC. The UMT uses the BDOC's secure communication equipment to coordinate with higher echelons and to coordinate area religious support requirements. (See Figure 6-4.)

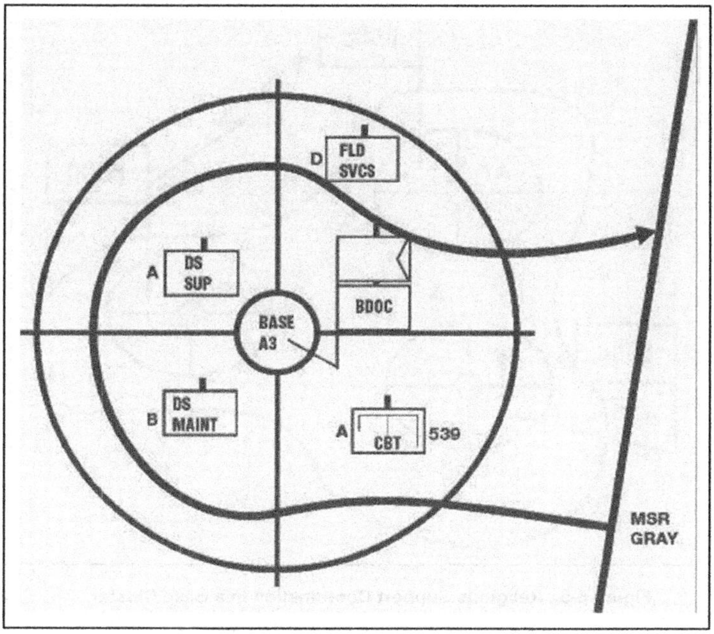

Figure 6-4. Religious Support Coordination in a Base

Base Clusters

6-46. A base cluster is a grouping of bases that are mutually supporting for defense purposes. The commander forms a Base Cluster Operations Center (BCOC) from his own staff and available base assets.

6-47. As in the base, the base cluster commander's UMT uses the BCOC as its source of information for coordinating area religious support in the base cluster and for communicating with higher and lower echelons. (See Figure 6-5.)

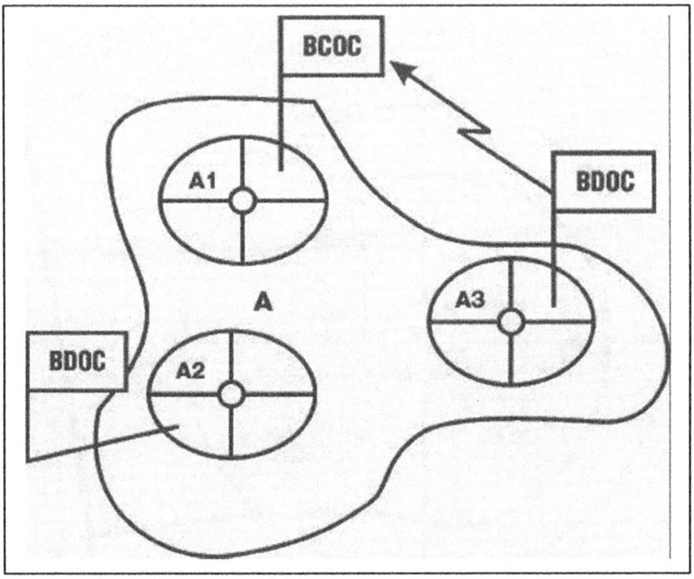

Figure 6-5. Religious Support Coordination in a Base Cluster

Religious Support Coordination In Rear Areas

6-48. The base or base cluster commander's unit ministry team (Base or Base Cluster UMT), regardless of the chaplain's rank, is responsible for area support planning and coordination. Tenant and transient units coordinate with the Base UMT. Base and Base Cluster UMTs coordinate scarce resources for optimum religious support.

6-49. Base and Base Cluster UMTs monitor unit movements, particularly company-size and smaller. They maintain a database of chaplains, their units, faith groups, locations, and means of communication.

6-50. When the commander does not have an assigned UMT, the senior chaplain in the base or base cluster coordinates area support. If there is no religious support available in the base or base cluster, the next higher echelon takes responsibility for area coordination. These relationships are specified in the Religious Support Appendix of the next higher headquarters.

Base Religious Support

6-51. Tenant units coordinate with the Base UMT for additional religious support. The Base UMT conducts a daily meeting to coordinate religious support for the soldiers in the base. Units without organic religious support contact the Base UMT through the BDOC for religious support.

6-52. The Base UMT reports tenant and transient chaplains by faith group to the Base Cluster UMT. This information is used to coordinate faith group support in the base cluster. Unit ministry teams provide area support as needed on a temporary basis until their unit's mission requires their movement from the base.

6-53. If there is no chaplain available in the base to meet a unique faith group requirement, the Base UMT requests support through the BDOC to the BCOC. For instance, if there is no Roman Catholic chaplain available, the Base UMT may request Catholic support using this procedure.

Base Cluster Support

6-54. The base cluster religious support is coordinated in the same way as in the base. The Base Cluster UMT monitors the movement of units into and out of the bases. As bases submit requests for additional faith group support, the Base Cluster UMT meets those requests with available UMTs. If the request cannot be met, it is forwarded to the next higher echelon. The Base Cluster UMT reports chaplains by faith group through the BCOC to the RAOC of the next higher echelon.

Higher Echelon Support

6-55. The procedure for coordination of faith group support described above is repeated up the chain of command. At higher echelons, as the area and the number of units increase, the maintenance of the chaplain database becomes more sophisticated. At these levels, a computerized database, using standard data base software, is required.

SPECIAL RELIGIOUS SUPPORT REQUIREMENTS IN REAR AREAS

RECEPTION, STATIONING, ONWARD MOVEMENT, AND INTEGRATION OPERATIONS

6-56. RSOI operations include the initial reception of units and individuals, the preparation of these units and individuals for combat, and their movement forward into the combat zone. Reception operations begin before hostilities start and continue after hostilities cease. RSOI may be conducted either forward in the theater of operation, in the rear of the area of operation at an Intermediate Staging Base (ISB), or at home station and/or en route to the AO. In the case of RSOI being conducted in the theater, the theater chaplain is responsible for coordinating religious support in the reception area. In the case of a RSOI being conducted closer to the AO, the COSCOM chaplain may be designated to provide area religious support.

MORTUARY AFFAIRS

6-57. The recovery, preparation, and transportation of the dead are one of the most difficult missions in the battlespace. Religious support is provided to soldiers in Mortuary Affairs units and to those detailed to perform this mission. The care for the dead requires sensitive religious support.

6-58. Planning Considerations:

- Religious support in sites where those killed in action are prepared for return to CONUS.

- Religious support for soldiers who are doing the work.
- Special faith group requirements for burial.

RECONSTITUTION AND REDEPLOYMENT

6-59. Reconstitution (both reorganization and regeneration) restores combat ineffective units to a specified level of effectiveness. The Army Service Component Command (ASCC) plans and conducts operational and tactical reconstitution operations.

6-60. Regeneration involves large-scale replacement of personnel and equipment and the rebuilding of units. This mission is usually assigned to the TSC, COSCOM, or DISCOM, depending on the size of the unit. Division, Corps, or ASCC UMTs assist in religious support to the units.

6-61. One or more UMTs may be assigned to a RTF to provide religious support to soldiers until the unit's assigned UMT has recuperated or been replaced. Religious support during reconstitution includes the ministry of presence, group and individual counseling and debriefing (see Appendix C, Critical Intervention Management and Critical Event Debriefings), worship and memorial services, and prayer. RTF UMTs play an important role in restoring the soldiers' spiritual fitness and rebuilding the unit's cohesion and morale.

6-62. Planning Considerations:

- RTF planning process and status.
- How many UMTs are required?
- Which UMTs are designated to join the RTF, how are they notified, and by whom?
- Anticipated duration of mission.
- Criteria for determining when RTF UMTs end the mission and return to their parent units.
- Logistical support for RTF UMTs.

THEATER RELIGIOUS SUPPORT

6-63. The Army Service Component Command (ASCC) serves as the senior army echelon in a theater and is the service component command of a unified command. It includes the service component commander and all Army personnel, organizations, units, and installations. The ASCC chaplain is the senior Army chaplain in a theater of operations and provides staff supervision over all Army religious support in the theater. The ASCC chaplain is responsible for recommending religious support policy to the ASCC commander. An example of an ASCC is Third U.S. Army, which is the ASCC of USCENTCOM.

OPERATIONS

6-64. The ability to provide religious support in the 21st century battlespace will depend upon five critical capabilities requirements: information access, situational awareness, communication, connectivity, mobility, and force structure. A deficit in any of these five requirements will radically impair the ability to provide timely religious support.

6-65. EAD UMTs must have real time operational awareness. They must be able to determine an appropriate response and expeditiously communicate that information to religious support personnel who will respond directly to the situation. To enable religious support throughout the battlespace, UMTs must be able to move and communicate seamlessly throughout the AO.

6-66. The complexity of coordinating religious support at EAD requires sufficient religious support personnel to provide flexible tailoring and augmentation to support the mission. The senior theater UMT estimates and plans campaigns with the staff and must focus on the different phases, determine priorities, and manage the coordination and execution of religious support tasks. When the theater commander establishes the combat zone (CZ) and communications zone (COMMZ), the theater UMT plans for simultaneous

6-66. activities across the full spectrum of operations throughout both AOs. Communication across both areas is critical for religious support synchronization; reach capabilities, reporting, and comprehensive planning. Division rear operations and Corps MSR operations are linked as a part of the battlespace management system, which includes tracking religious support assets from COMMZ through assembly areas (AA) to corps rear to division.

STRATEGIC RELIGIOUS SUPPORT PLANNING

6-67. The combatant commander's UMT's strategic religious support planning provides the framework for the religious support assets in peacetime and in crisis. During predeployment, the senior command UMT develops an assessment that transitions the focus to crisis or war. Force projection and assessment transitions religious support assets to combat operations. Religious support planning is integrated into the combatant commander's staff plans and estimation process. The result is expressed in terms of religious support to mission objectives, concepts, vision, resources, and provides guidance to subordinate command UMTs for a broad range of religious support activities. Planning should consider—

- The command's desired end state and how religious support can support full accomplishment of military objectives.
- Risks associated with METT-TC and the religious support assets.
- Religious support courses of action meeting mission objectives.
- Visualization of the religious support needs.

STRATEGIC RS ESTIMATE

- Assigned objectives from combatant commanders.
- Translation of combatant commanders objectives applicable to theater RS.
- Visualization of the strategic environment; how it relates to the accomplishment of the commander's RS objectives.
- Assessment of the threats of accomplishment of assigned RS objectives.
- Assessment of strategic alternatives to available, with accompanying analysis, risks, and the requirements for RS plans.
- Impact of indigenous religions on operations.
- Considerations of available RS resources, linked to accomplishment of assigned RS objectives.

ARMY SERVICE COMPONENT COMMAND RELIGIOUS SUPPORT

6-68. The ASCC chaplain manages UMT, chaplain and chaplain assistant, assignments throughout the theater for the commander. The senior chaplain assistant NCOIC advises the senior chaplain on all chaplain assistant assignments. The ASCC command UMT provides training, policy, logistics, and funding for all Army UMTs in the theater and serves as primary liaison with the next higher commander's staff chaplain. The ASCC UMT stays in close communication with senior UMTs in each sector. They participate in joint planning to ensure that the interest of the commander's religious support goals and plan are represented and implemented. They coordinate with the TSC UMT for movement, and transportation unit information and resources. The Personnel Command (PERSCOM) UMT in theater coordinates assignments for UMTs for the TSC command UMT. Functions of the ASCC UMT include—

- Establish liaison with host nations religious leaders and other Joint Religious Ministry Support Teams at the highest levels in the theater.
- Coordinate fund transitions for logistical support.
- Coordinate religious support for force arrival in transfer of authority (TOA) and RSOI sites.
- Coordinate religious support for contractors for the command as part of the commander's policy.
- Provide guidance to all senior Army UMTs to support the combatant commander's RSP.
- Plan and provide training for UMTs.
- Manage UMT assignments within the theater to ensure appropriate denominational balance to all areas of the theater.
- Provide guidance for the employment of all Chaplain Detachments (CDs) and SMART-PC in support of the religious support mission.

CHAPLAIN DETACHMENTS

6-69. When requested from a MACOM, the HQDA G3 will task USARC through FORSCOM to mobilize and deploy the requested chaplain detachments needed to support planned, current, or ongoing operations. The Chief of Chaplains may request that HQDA G3 task USARC to mobilize and deploy chaplain detachments in support of specific missions. (See Appendix G.)

ADDITIONAL CORPS/ECHELONS ABOVE CORPS RELIGIOUS PLAN REQUIREMENTS

SPECIAL OPERATIONS

6-70. The mission, duration and scope of operations, security considerations, signal capabilities, and the desired degree of control determine Special Operations Force (SOF) command relationships. SOF units above group or regimental level are Joint. Joint force commanders include unified, subordinate unified, JTF, functional component, and commanders of specific operational forces. USSOCOM organizes, equips, trains, and provides Special Operations Force for combatant commanders of regionally oriented unified commands (European, Atlantic, Southern, Pacific, and Central) and for the Commanding General, U.S. Forces, Korea. Combatant commanders delegate operational control of these SOF forces to their theater-level Special Operations Commands (SOCOMs). The six SOCOMs are the focal point for in-theater SOF, form nuclei for the Joint Special Operations Task Force (JSOTF), and furnish expertise needed to effectively employ SOF independently or in concert with conventional forces.

6-71. A unified command normally consists of the four service-specific conventional forces and a SOCOM. The combatant commander or the SOCOM may establish a JSOTF commanded by the Joint Force Special Operations Component Commander (JFSOCC). SOF units from the three services are normally under the OPCON of the JFSOCC.

6-72. ARSOF missions are often highly classified and sensitive. To provide counseling, debriefing, and other religious support to ARSOF units, the chaplain must have a Top Secret clearance. Chaplain assistants are required to have a clearance equal to mission classification. The USASOC UMT recommends all ARSOF UMT assignments.

6-73. Normally, the JSOTF chaplain is a member of the service having the majority of forces. (See JP 1-05, Chapter 2.) If the Army supplies the majority of forces, an ARSOF chaplain is normally designated as the JSOTF chaplain.

6-74. The JSOTF chaplain coordinates with the senior chaplain of the next higher headquarters to ensure complete religious support for SOF units. The JSOTF chaplain, in turn, supervises SOF religious support throughout the theater. The chaplain coordinates for additional religious support when other SOF chaplains are unable to provide religious support. SOF chaplains may provide limited area support for conventional forces when it does not conflict with SOF mission requirements

MEDICAL COMMAND SPECIAL MEDICAL AUGMENTATION RESPONSE TEAMS-PASTORAL CARE

6-75. The U.S. Army Surgeon General (DASG) and the U.S. Army Chief of Chaplains (DACH), in consultation with the Chief of Staff of the Army, the Director of Military Support, and applicable Federal and Department of Defense regulations, requires designated DASG/MEDCOM/DACH assigned, attached or associated individuals and organizations to develop and maintain special organized, trained and equipped SMART-PC.

6-76. Operations consist of—

- Alert, assemble, and issue warning order within 6 hours of no-notice notification.
- Deploy from home station within 6 hours of the warning order.
- Conduct mission: supporting commander without organic ministry team or needing additional ministry team support for mass casualty or devastating contingency operation; or provide professional pastoral care augmentation (to include Critical Event Debriefing, Trauma Ministry, Mass Casualty Ministry, and Spiritual Assessment) to local authorities in the management of disaster/mass casualty incidents.

6-77. SMART-PCs deploy with enough food, clothing, personal hygiene supplies, and equipment for self-sustainment in an austere environment for 72 hours. Team missions normally conclude in 72 hours. When extenuating or mitigating circumstances require DASG/MEDCOM or DACH approval for mission extension, SMART-PC members will procure locally or formally obtain logistical support from supported units, or local civil or Lead Federal Agency. SMART-PC members deploy on orders articulated on DD Form 1610.

6-78. At the direction of DACH and/or on order of DASG/MEDCOM, at request of commanders, and/or the request of legitimate civil, Federal, or Defense authorities, using appropriate, recognized, and approved channels designated units will deploy SMART-PC to provide short duration, medical ministry augmentation to:

- Regional, domestic, Federal and Defense agencies responding to disaster, civil-military cooperative action, humanitarian and emergency incidents.
- Support units without organic ministry teams or who require additional ministry team support for mass casualty or devastating contingency operation.

6-79. SMART-PC is an addition to the list of required teams, which includes:

- Trauma/Critical Care (SMART-TCC).
- Chemical/Biological (SMART-CB).
- Stress Management (SMART-SM). (This team has one CH assigned, no assistant.)
- Medical Command, Control, Communications, Telemedicine (SMART-MC3T).
- Preventive Medicine/Disease Surveillance (SMART-PM).
- Burn (SMART-B).
- Veterinary (SMART-V).
- Health Systems Assessment and Assistance (SMART-HS).

6-80. SMART-PC is modeled on the standard Unit Ministry Team of one chaplain and one chaplain assistant. Operationally, multiple teams are task organized based upon METT-TC and medical-ministry/risk analysis to provide the appropriate level of response and technical augmentation to meet mission needs.

MANAGEMENT:

DASG, ASSISTANT SURGEON GENERAL, FORCE PROJECTION

Policy, supervision, and oversight to equip, deploy and redeploy SMART-PC. Validate and resource special equipment requirements; periodically test and evaluate individual SMART-PC teams; develop, monitor, and report readiness reporting requirements. Establish and maintain officer liaison with DACH.

DASG / MEDCOM, ASSIGNED REGIONAL MEDICAL COMMANDS

Organize, train, and equip 2-SMART-PC teams. On order, deploy assigned SMART-PC within assigned regional medical boundaries. Be prepared to deploy outside of regional boundaries in support of other SMART-PC or to other national/international sites.

EQUIPMENT: Configuration of SMART-PC deployment sets, kits, and outfits will be man portable, and utilize commercial backpacks, suitcases, and rugged containers. Each equipment piece will meet commercial airframe size, weight, and cube requirements.

MEDICAL COMMAND SPECIAL MEDICAL AUGMENTATION RESPONSE TEAM-STRESS MANAGEMENT

6-81. The role of the chaplain assigned to a SMART-SM is to bring spiritual aspects of stress management, as a clinician, in support of the team's overall tasks, which include:

- Critical event debriefings.
- Spiritual counseling.

CONTINGENCY FORCE POOL (CFP)

6-82. The FORSCOM Chaplain's Office will coordinate with CONUSA Chaplains where CFP chaplains are attached for duty to ensure that chaplain personnel are integrated into MSCA contingency training and planning scenarios, and are included in contingency plans. The FORSCOM Chaplain's office will coordinate with the CONUSA Chaplains to de-conflict training and planning objectives. FORSCOM will ensure that CFP personnel support is placed in the Regional Planning Agents (RPA) MSCA handbook, in the DOD database to the Defense Coordinating Office (DCO) and integrate CFP training / certification with DOD Emergency Preparedness Training and CONUSA MSCA contingency exercises. Upon request, FORSCOM has the ability to cut orders for CFP personnel using ARPERCOM Orders and Resource System (AORS) and can make one or more CFP personnel immediately available depending on the crisis level. In a small-scale contingency operation (SSCO) OCONUS response, CFP personnel will report to the CONUS MACOM for contingency support, providing religious support and liaison / coordination. During MSCA or Homeland Defense / Security (HLD / HLS) CONUS contingencies, upon request by a DCO and with appropriate funding for RS to DOD personnel, CFP personnel become assets to the JTF Chaplain.

ROLE OF THE CORPS/ECHELON ABOVE CORPS UNIT MINISTRY TEAMS

6-83. The personal delivery of religious support by the UMT will be imperative in all operations. It is essential that the Corps/EAC HQs UMT is able to rapidly tailor and modularize UMTs for operations. Corps/EAC UMTs must organize to optimize and use information. Flexibility in applying METT-TC is critical to accommodate specific missions. Rank structure and composition of UMTs may change to support specific missions. RS planning must be robust and flexible enough to provide religious support across the full spectrum of military operations. Corps/EAC UMTs require the same level of technology as the supported force. As technology advances the UMT automation, communication and transportation requirements must keep pace with the command's overall requirements.

GARRISON OPERATIONS

6-84. During garrison operations, all EAC UMTs, the Corps UMT, and subordinate UMTs support the training program of the installation chaplain. They implement their CMRP to support the religious needs of the families and soldiers. The EAD UMTs manages and supervises subordinate UMTs assigned for the commander. They support the commander's training program through participation and planning and implementing UMT training. The Corps UMT is a mentor to the BDE UMTs and will execute a mentorship program. The Corps UMT supports the ASCC UMT RS plan.

SUMMARY

6-85. Corps and EAC UMT role is crucial to the warfighting religious support mission success. AC and RC are fully integrated to supply combat ready UMTs to provide religious support to the full spectrum force that meets the needs of the joint force commander in war, conflict and peace. Command UMTs are fully trained, ready, and staff integrated to meet the religious support needs of soldiers, family members, and authorized civilians in all operations. The UMTs are strategically and operationally responsive and prepared to conduct prompt and sustained religious support operations as part of joint, multinational, and interagency teams.

RESOURCES

6-86. The following is a list of main resources utilized for writing and updating this doctrine.

- AR 165-1, Chaplain Activities in the U.S. Army
- JP 1-05, Religious Ministry Support for Joint Operations
- JP 5-00.2, Joint Task Force Planning Guidance and Procedures
- JP 3-05, Doctrine for Joint Special Operations
- JP 3-0, Doctrine for Joint Operations
- JP 3-07.3, Joint Tactics, Techniques, and Procedures for Peace Operations
- JP 1-0, Doctrine for Personnel Support to Joint Operations
- JP 3-08, Interagency Coordination During Joint Operations

- JP 3-10, Joint Doctrine for Rear Area Operations
- FM 1 (100-1), The Army
- FM 16-1, May 1995, Religious Support
- FM 3-0 (100-5), Operations
- FM 25-100, Training the Force
- FM 25-101, Battle Focused Training
- FM 100-15, 29 OCT 1996, Corps Operations
- FM 100-11, 15 JAN 1998, Force Integration
- FM 100-9, 13 JAN 1992, Reconstitution
- FM 100-8, 24 NOV 1997, The Army In Multinational Operations
- FM 100-7, 31 MAY 1995, Decisive Force: The Army in Theater Operations
- FM 100-6, 27 AUG 1996, Information Operations
- FM 22-100, Army Leadership
- DA Pam 165-3, Chaplaincy Training Strategy
- National Military Strategy
- Joint Vision 2010
- Joint Vision 2020
- Army Vision 2010
- Concept for Future Joint Operations, May 1997
- U.S. Army Chaplaincy Strategic Plan FY 2000-FY 2005
- Chaplain Support Force XXI White Paper
- Lessons Learned from past and current operation
- Lessons Learned from the training centers

Chapter 7

INSTALLATION UNIT MINISTRY TEAM RELIGIOUS SUPPORT OPERATIONS

7-1. The Army, Active and Reserve components, is designed, equipped, and trained for force projection operations. Army installations function as force projection platforms, providing support for deployed tactical units throughout all stages of force projection. To ensure readiness, the Army continually plans and trains for the full range of military operations.

CONTENTS

7-2. Combat readiness also produces an Army whose organization, skills, leadership, and discipline can advance U.S. interests in a variety of operations not characterized as war. These operations include, but are not limited to, nation assistance; humanitarian and disaster relief; counter drug operations; peace operations; antiterrorism; shows of force; attacks and raids; noncombatant evacuation operations; insurgency and counterinsurgency support; civil disturbance and support to domestic authorities.

7-3. Force projection is the military component of power projection and is a critical component of U.S. strategy. AC and RC units, the mobilization base, DA civilians, and the industry are directly involved projecting the force anywhere in the world. Training, mobilization, deployment, employment, sustainment, and redeployment are the processes utilized for force projection. They are continuous, overlapping, and interactive throughout an operation. The process of force projection is an integral part of the Joint Operation Planning and Execution System (JOPES). JOPES is used to monitor, plan, and execute mobilization, deployment, employment, sustainment, and redeployment activities associated with joint operations. The Army force projection process is imbedded within this system.

7-4. Religious support is a vital part of the combat readiness mission of the installation in force projection. Force projection operations are inherently joint and religious support planning and synchronization are required.

INSTALLATION DESCRIPTION

MISSION

7-5. Army installations, which include STARCs and Reserve Support Command (RSCs), contribute to the Army's ability to support the National Military Strategy (NMS). Peacetime missions include: counter-drug operations on our national borders, nation building, counter-terrorism, WMD support, deployment of land forces for a conflict, or fighting and winning the nation's war. Each installation, STARC, and RSC has a primary mission that determines its assignment to a MACOM or Component.

ORGANIZATION

7-6. Installations may host dozens of organizations that are either part of the MACOM's mission or simply tenants assigned to the installation for support. Installations are organizationally characterized by the nature of the major activity they host. For force basing they would have Corps and Division(s). For unit training they would host such units, as Combat Training Centers (CTC), for example, the Joint Readiness Training Center (JRTC). For the logistic support they would provide depot and commodity management. Within the Army, an installation may be referred to as a post, camp, station, fort, barracks, subpost, depot, arsenal, proving ground, base, laboratory, port, or ammunition plant. For the Army National Guard and Reserve Component the term "installation" refers to the RSC for the RC and the STARC for the ARNG.

As Part Of A Major Army Command

7-7. Army installations are assigned to and operate under commanders of Major Army Commands (MACOMs) or MSCs where applicable, such as in the Army Materiel Command (AMC), and the U.S. Army Reserve Command (USARC). STARCs operate under the guidance of the National Guard Bureau. Installations consume, provide, maintain, and otherwise control significant resources while executing DOD, DA, and MACOM mandated policy programs. These programs are reviewed, analyzed, and resourced at DA level. Transformation Installation Management (TIM) will eventually transform the relationships between installations and MACOMs, to include, the relationships between MACOM Chaplains and Installation Chaplains. UMTs should keep abreast of such changes.

As Part Of Full Spectrum Operations

7-8. The entire installation management system is a key component of developing and sustaining a total force capable of meeting the full spectrum of missions required by the National Military Strategy (NMS). The Army's mission to expand and rapidly deploy trained and ready units is dependent upon installation capabilities. Within the installation support system is the Army's mobility infrastructure. Total force readiness is linked to installation operations.

KEY TERMS

- Joint Operation Planning Execution System: (JOPES)
- Area Support Group: (ASG)
- Port of Embarkation: (POE)
- Arial Port of Embarkation: (APOE)
- Seaport of Embarkation: (SPOE)
- Port of Debarkation: (POD)
- Aerial Port of Debarkation: (APOD)
- Seaport of Debarkation: (SPOD)

7-9. Army installations are standardized to ensure that operations can be effectively supported in peacetime. This effectiveness must continue throughout the transition from peacetime to wartime (mobilization) and during post conflict (post-deployment). Army installations provide the sustaining support necessary for units to accomplish their missions.

7-10. Installations provide this support by accomplishing both their primary and support missions. The primary mission of the installation includes, but is not limited to, readiness and preparation for operational deployment, training, supply, and/or maintenance. The support mission includes real property acquisition and upkeep, community service, and other personnel support.

7-11. Sustainment includes the primary and support missions of the installation. Throughout each phase of deployment, the installation manages the sustainment of units.

FUNDAMENTALS OF INSTALLATION RELIGIOUS SUPPORT OPERATIONS

7-12. Military chaplains and chaplain assistants are specially trained and assigned to provide for the government's interest to protect free exercise of religion rights in the military, which is a unique pluralistic and multi-faith environment. Military chaplains and chaplain assistants assigned in Table of Distribution and Allowances (TDA) positions provide direct support to the military commanders for the installation religious program, Family Life Centers, units, hospitals, initial entry training (IET) units, Chaplaincy resources management, Chaplaincy personnel management, Chaplaincy force structure and training. The content of the religious support mission of the chaplain and chaplain assistant is no different in the TDA than the TOE; only the command context changes. It must be understood that the Chaplaincy is not a social service under the city management context of the Installation, but a direct support function to the commander. Religious support is the military commander's mission. TDA positions are required to develop military religious support leaders and train religious support skills for the full spectrum of military operations, from the sustaining base to the forward edge of the battlespace.

7-13. Installation religious support has three major areas:

- To prepare and assist soldiers and families for military operations.
- To provide religious support for units and personnel on the installation.
- To sustain religious support operations during deployments and mobilizations.
- To train UMTs for comprehensive RS.

7-14. The installation chaplain is responsible for all religious support on the installation including all religious services and activities. The installation chaplain coordinates with tenant units to implement the CMRP. The installation's UMTs provide a broad religious support programs for units, soldiers, and their families. Worship services, pastoral care, religious education, and spiritual fitness training are provided for the religious support needs of the installation population.

READINESS

7-15. Military chaplains and chaplain assistants in the TDA are necessary to support peacetime deployments, as well as, to man operational units in mobilization and provide wartime rotation, and rapid casualty replacement. All TDA military chaplain and chaplain assistant positions are required to be trained and ready to use combat skills and training to implement the military religious support mission of the commander in any contingency.

7-16. The Reserve Component (RC) holds a significant portion of the mobilization assets for the Chaplaincy. The RC Chaplaincy is a military combat augmentation in direct support of the commander. RC chaplains and chaplain assistants in the TDA are necessary to support peacetime deployments, as well as, to man operational units in mobilization, provide wartime rotation and rapid casualty replacement. All RC TDA chaplains and chaplain assistants are required to be trained and ready to use combat skills and training to implement the religious support mission of the commander in any contingency. The RC Chaplaincy provides the recruitment base for both the AC and RC. The RC provides the training base for the Chaplain Candidate Program. The IMA and Active Guard Reserve (AGR) chaplains and chaplain assistants are in the TDA structure.

Assessment of Unit Ministry Team Readiness

7-17. As the commander's personal staff officer for religious support, the command installation chaplain is responsible for assessing the technical proficiency of UMTs in subordinate and tenant units. They conduct assessments utilizing the units training program such as during field training exercises (FTXs), external Army Training and Evaluation Program (ARTEP) events, unit training, Reserve component annual training and Command Inspections. Observer controllers may provide additional assessment assistance when the UMT is involved in Battle Command Training Program (BCTP) events or Combat Training Center (CTC) rotations. They acquire, in coordination with the senior unit UMT lessons learned to integrate into the installation UMT training program. (See DA PAM 165-3, The Chaplain Training Strategy.)

7-18. Chaplains and chaplain assistants participate in the installation's Command Master Religious Plan (CMRP) to help soldiers develop the personal spiritual readiness needed to sustain them during combat. Spiritual readiness is also essential to sustain soldier families while the unit is deployed. As part of the installation's CMRP, the UMTs conduct worship services, offer religious education, and provide spiritual fitness training and other activities in installation chapels and facilities.

Training

7-19. Training is the cornerstone of preparation for military operations. The Army's primary activity during peacetime is realistic, battle-focused training oriented on the unit's mission essential task list (METL). Installation chaplains are responsible for individual training, MOS 56A and 56M. The installation chaplain's training managers coordinate with subordinate UMTs and unit S3/G3s and ensure that UMTs have adequate technical and tactical training for deployment and combat. This is implemented through the development and execution of an annual Installation wide UMT training strategy and plan (the religious support METL provides training tasks along with the Chaplain Officer and Enlisted Critical Task Lists). This training is integrated within the implementation of the Chief of Chaplain Training Strategy (DA Pam 165-3) and higher headquarters training strategy.

7-20. Tactical proficiency is a demonstrated understanding of the Army's war-fighting doctrine and tactics. To provide effective religious support in combat, the UMT must know its unit's doctrine, tactics, techniques, and procedures. The UMT must also maintain the requisite skills of field craft essential for survival in the battlespace. Tactical training is primarily the responsibility of the unit to which the UMT is assigned or attached.

7-21. Technical proficiency is the demonstrated ability to perform all tasks related to one's functional area. Technical proficiency is required for both garrison and field missions. Technical proficiency is acquired from schools, self-study and training with other UMTs in performing the unit's and installation's religious support mission. Training and continuing education conferences, functional courses offered by USACHCS, and courses offered through the AMEDD Center and School, and the NCO Education System (NCOES) supplement the installation religious support training program.

7-22. UMTs must balance religious support training for combat and for garrison. The UMT must participate fully in both unit and religious support training to assure their tactical and technical proficiency. If training for either combat or garrison operations is neglected, ministry teams will not be prepared to provide comprehensive religious support to soldiers and their families.

7-23. Many installations serve as mobilization bases. (See Figure 7-1.) Mobilization can overwhelm the installation's resources. Thorough planning and realistic training reduces confusion. Mobilization training gives UMTs an opportunity to practice their mobilization religious support missions and an opportunity to assess the adequacy of mobilization plans. Installations assist with the following RC UMT training events:

- Annual training (AT).
- The Individual Mobilization Augmentee (IMA) training.
- Individual training events, Individual Ready Reserve (IRR), and chaplain candidate training.
- AC/RC Sustainment Training.
- Annual Chaplaincy Regional Sustainment Training (CREST) as mandated by the Chief of Chaplains.

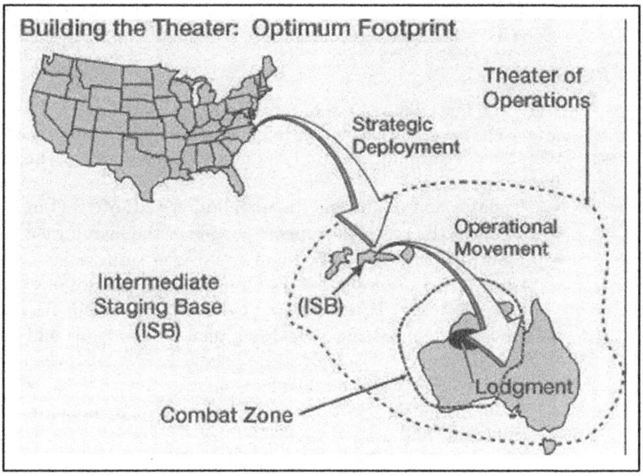

Figure 7-1. Building the Theater

FORCE PROJECTION

7-24. When the installation is serving as a force projection platform, the installation UMT must be prepared to provide as a staff the religious support mission. The installation chaplain's staff must be prepared to support all phases and sustainment requirements of the mission. Provision is made for religious support for all phases of deployment including pre-deployment, movement to the POE, strategic lift, theater reception, and theater onward movement. This includes counseling, training of UMTs, leadership training, reunion briefings and counseling services (and as a family readiness group resource). Installation UMTs will support peacetime deployments, as well as, to man operational units in mobilization, provide wartime rotation and rapid casualty replacement.

7-25. Force Projection planning includes—

- Mobilization, receiving, housing, processing, validating, deploying of soldiers including UMTs.
- Support the increased information communication requirements that will occur.
- Assist in reception as the force shifts MRC priority and re-deployment of forces.
- Maintain installation religious support requirements for the remaining military community after deployment.
- Training of UMTs as required.

MOBILIZATION PHASES

(See CH 3 for more information.)

PLANNING PHASE

7-26. The installation chaplain becomes the mobilization station chaplain and assumes a key role when RC UMTs mobilize. Before mobilization, the installation chaplain designates a mobilization planner who does the following:

- Updates and maintains the installation's RC MOB Plan book.
- Updates the religious support portion of the installation MOB PLAN.
- Identifies positions to be filled by IMAs or retirees.
- Clarifies cross-leveling responsibilities and coordinates with the G1/Director for Personnel and Civilian Affairs (DPCA).
- Develops a mobilization-training plan for chaplains and chaplain assistants.
- Prepares UMTs for deployment.
- Prepares for and participates in installation mobilization exercises (MOBEX) and activities.
- Plans for demobilization activities such as after action reviews, reunion briefings, awards, reports, etc.
- Maintains and improves combat readiness posture.
- Ensures all sites, such as RSOI, are fully manned by UMTs during all phases of mobilization, deployment, and redeployment.
- Family Readiness Group (FRG) religious support.
- Prepare the UMT MOBTDA.
- Coordinates with and tracks RC units with and without UMTs scheduled to report to the installation.
- Communicates with IMAs and retirees with orders to report to the installation upon mobilization.

7-27. This phase concerns all AC and RC UMTs efforts at all levels during peacetime to plan and train for assigned wartime tasks. The installation command UMT should ensure that the installation RS mobilization plan addresses civilian support for the mission. They must review the plan annually ensuring it is up-to-date and fully integrated into the installation commander's mobilization plan.

7-28. STARC UMT mobilization plans and SOPs clarify training and MOBTDA personnel to provide religious support to families at family assistance centers (FACs). Important consideration must be given to identifying the religious support requirements of the FACs and developing MOBTDAs with sufficient UMT personnel support.

7-29. The CONUSA UMT coordinates between the FORSCOM staff UMT and RC UMTs within its geographic area. CONUSA staff chaplains and NCOICs may redistribute UMT members in order to match anticipated UMT mobilization requirements.

ALERT PHASE

7-30. This phase begins when an AC or RC unit receives notice of a pending mobilization order. During this phase, the STARC and Army Reserve Command chaplains and NCOICs continue to cross-level individual members of UMTs within their state or region to bring alerted units to required religious support deployable status.

HOME STATION PHASE

7-31. This phase begins with the RC unit's entry on active federal duty or the AC's preparation for deployment. Actions are preformed to bring units to active status. Installation, STARC, and USARC Command UMTs continue cross-leveling, as well as, ensuring critical UMT logistical support is in place.

MOBILIZATION STATION PHASE

7-32. This phase begins when the AC/RC unit arrives at the MS or mobilization site. The unit UMT is validated to meet deployment unit validation criteria resulting in assuring religious support mission capability. Necessary chaplain and chaplain assistant individual and collective training are conducted. Timeliness is critical. Integration to the Soldier Readiness Program (SRP) for preparation of overseas movement (POM) activities for religious support input is conducted.

PORT OF EMBARKATION PHASE

7-33. This phase begins with the arrival at the APOE or SPOE. Religious support is critical to ensure the forces are spiritually prepared for deployment and combat. Worship, counseling, and other religious support tasks are initiated. UMTs departing are ensured spiritual support.

DEPLOYMENT PHASE

7-34. In addition to routine religious support operations, the installation's TDA UMTs have the responsibility to plan and to prepare to sustain religious support operations on the Installation throughout the mobilization, deployment, and redeployment and demobilization process.

7-35. The senior UMTs of the deploying units and the TDA UMTs coordinate plans for religious support during deployment. Planning considerations include the following:

- The care of families and authorized civilians.
- Procedures for consolidating chapel programs.
- The close-out of chapels (if necessary).
- The accounting, turning-in, or securing of unit and installation property.
- Survivor and casualty assistance.

7-36. FORSCOM Regulation 500-3-2 (FORMDEPS II), Deployment Guide, provides guidance and assigns responsibilities for support of the deployment of units.

7-37. UMTs of non-deployed units are included in installation plans to support deploying units. A garrison support unit (GSU) may be available to augment the installation's UMTs during deployment and mobilization.

Along with the installation's non-deployed UMTs, they assist with religious support to families until the deployed forces return.

7-38. A thorough understanding and appreciation for the Transfer of Authority of deploying units to the theater commander is critical to the installation chaplain's mission. At a time designated in the deployment order, the chain of command shifts to the gaining unit, usually the theater commander. When that TOA occurs, the relationship of the deploying UMTs to the installation's UMTs changes correspondingly. This new relationship remains in effect until the unit redeploys.

SUSTAINMENT PHASE

7-39. To sustain the force projection platform, the installation UMT planner must consider all sustainment requirements for the non-deployed units, units at all deployment stages, and family members of both military and authorized civilians. The continued training and readiness posture is maintained for sustaining the forward religious support mission.

Sustainment: Family Readiness Groups

7-40. Command sponsored Family Readiness Groups (FRGs) are a vital link between soldiers, their families, the command, and the unit. Volunteer family members lead FRGs. Each deploying unit establishes its own FRG. Establishing the FRG prior to the deployment alert is extremely helpful in developing support plans and operational procedures for providing assistance to the unit' families. Chaplains and chaplain assistants support these groups.

Sustainment: Casualty Notification

7-41. Notification of the next of kin (NOK) is the commander's responsibility. Chaplains are not notification officers, but they may be a part of the team that makes the notification. In the event of a disaster or mass casualty situation, they may serve on the Casualty Assistance Team. Religious support to grieving families is a priority for UMTs and appropriate training in such religious support is a necessity.

REDEPLOYMENT AND DEMOBILIZATION PHASE

7-42. Religious support during the redeployment and demobilization phases consists of the spiritual and emotional preparation of units from an area of operation to follow-on designated CONUS or OCONUS bases. Redeployment religious support must be planned and executed to insure the proper allocation of resources of an installation. Planning must encompass all phases of an AC force returning to its home station, mobilized RC forces, and individuals through DMS back to their home stations.

7-43. This phase begins with the arrival of forces at the mobilization station/installation. The installation UMT supports the commander in receiving the forces, receiving sustainment equipment and supplies, and helps develop a reception and reconstitution or onward movement plan.

7-44. Demobilization is a critical factor for strategic reconstitution. The installation staff UMT plans and executes the commander's demobilization plan. Reunion support to soldiers, authorized civilians, and families are given to ease the transition from mobilization/deployment to demobilization/redeployment. Counseling, both individual and group, workshops, and worship are critical for reconstitution. Demobilization of some units may occur at the same time other units are being mobilized, deployed or redeployed. Religious support planning must encompass a possible ongoing planning and execution process.

7-45. Religious support plans encompass—

- Reunion briefing for family members and friends.
- Reunion support for returning units.
- Back briefing from returning and TDA UMTs.
- Religious support to garrison command and staff.
- Critical event debriefing.

DOMESTIC SUPPORT

7-46. 7-46. A presidential declaration of a major disaster or emergency usually precedes Domestic Support Operations (DSO). The installation staff UMT plans and executes the commander's domestic support mission. As in other operations, the UMT mission does not change. The installation becomes the force projection for Army forces to the area of operation. DOD Directive 3025.15, Military Assistance to Civil Authorities (MACA) provides directives and guidance in responding to natural and manmade disasters and includes military assistance to civil disturbance, counter drug activities, combating terrorism, and law enforcement. Installation UMTs must know their role and limitations in this type of operation. Installation UMTs support local, state, and federal civil authorities (see JP 3-07.7; JP 1-05; FM 100-20).

ROLE OF THE INSTALLATION UNIT MINISTRY TEAM

7-47. The Installation Chaplain and staff provide seamless religious support to the non-linear battlespace extending from the CONUS deployment and sustaining base to the forward edges of the battlespace. The religious support mission extends throughout the full spectrum of military operations. The community religious support mission encompasses all military units, activities, organizations, and authorized personnel.

7-48. The importance and influence of the Chaplaincy to the spiritual and moral health of the unit (to include the sustaining base or garrison) and in spiritual matters has been valued throughout the history of the Army. Commanders recognize the value of the chaplain and the total UMT both in combat and garrison.

RESPONSIBILITIES AND ACCOUNTABILITY

Staff Responsibilities

7-49. The Installation Chaplain provides support and advice to the commander and staff and to all installation agencies on all matters of religion, ethics/morals, and morale as affected by religion. The Installation Commander is responsible for the religious support and spiritual fitness of soldiers, family members, and authorized civilians. This includes insuring comprehensive worship opportunities that are part of the CMRP.

7-50. Army chaplains and chaplain assistants are the agents of trust representing the Army's values and the needs of the military in the commander's religious support program. Chaplains and chaplain assistants are required to provide religious support in all military units, activities, organizations, and communities.

CHAPLAIN (O-6)	***CHAPLAIN ASSISTANT NCO (E-7/ 8)***
Role Commissioned Officer/Staff Religious Professional Title 10 Protected	***Role*** Noncommissioned Officer Specialized Assistant Religious Support Specialist
Mission Advises commander on religious support and quality of life issues to support the unit's mission. Performs/provides/coordinates the religious support mission for the command. Installation Director of Religious Support	***Mission*** Advises Chaplain and NCO Chain. Provides specialized assistance in areas of religious support to support the chaplain. Perform Garrison Staff functions. Installation Religious Support NCOIC.
Parallel Tasks 1. Supervise, mentor and assess subordinate Chaplains. 2. Recommend personnel assignments. 3. Supervises the contingency, mobilization plans. 4. Responsible for religious support training and professional development.	***Parallel Tasks*** 1. Supervise, mentor and assess subordinate chaplain assistants. 2. Recommend Chaplain Assistant assignments. 3. Prepares the contingency, mobilization plans. 4. Plans and conducts religious support training.

7-51. Installation Chaplains have staff responsibility for the following:

- Development and execution of the Command Master Religious Program under the direction of the commander; which includes all worship services, memorial services, religious activities, professional training, religious education, spiritual fitness and moral leadership training, mobilization plans and contingency operation planning, and pastoral support for soldiers, families, and authorized civilians.
- Conduct of officer personnel management and coordination enlisted personnel actions for Unit Ministry Team (UMT) assignments and professional development.
- Development of an annual training plan for chaplain and chaplain assistant training.
- Development and execution of the installation commander's suicide prevention program.
- Chaplain Family Life Centers which provide comprehensive programs for training UMT members, and family programs, such as, marriage and family counseling and family enrichment programs.

- Administration and management of the installation non-appropriated chaplains' fund.
- Supervision of the use of chapels and religious facilities to include planning for new and renovated religious use facilities.
- The Installation Chaplain is the personal staff officer to the commander for the religious support portions of planning, programming, force structure actions, budget input, crisis management, and force protection.
- Provide the commander unit spiritual fitness assessments.
- Support the commander's METL through implementing a religious support installation METL.

7-52. The TDA force structure is required in the active and reserve component to maintain adequate military rotation and surge capability for mobilization and deployments and for military career progression of chaplains and chaplain assistants. These TDA positions are required to train and develop military religious support leaders in religious support skills for the full spectrum of military operations.

Relationship To The Chief Of Chaplains

7-53. The Installation Chaplain section implements the Chief of Chaplains (CCH) Strategic Plan and guidance on providing religious support across the full spectrum of operations from the power projection base to the theater of operations and return. They plan and execute the Chaplaincy mobilization and contingency plans. The installation staff UMT implements the CCH training strategy for individual UMT members and collectively. They train and prepare UMTs in support of the CCH guidance for the Army.

Relationship To Major Army Commands

7-54. Installations are part of Major Army Commands (MACOMs) or MSCs. The installation will coordinate, review, and recommend assignment of chaplains and chaplain assistants. They will support the MACOM or MSC staff chaplain contingency plans. Also, they will implement the religious program as set forth through the command. Again, as Transformation Installation Management takes effect, the role between Installation Chaplains and MACOMs will likely change. UMTs should keep abreast of such changes.

SUPERVISORY RESPONSIBILITIES

7-55. The assignment instructions received by every chaplain at the time of his/her receipt of permanent change of station (PCS) orders, contains the following statement: "Chaplain will perform and/or provide direct and general religious support in accordance with Unit Religious Support Plan and Command Master Religious Program." The installation staff UMT establishes staff supervision and implements the CMRP. They supervise the training of UMTs for the CCH in cooperation and coordination with tenant command UMTs. They supervise garrison religious support force structure for the commander, both active and civilian. On an installation or in a community, the staff chaplain has overall staff responsibility for the religious support mission. In fulfilling that responsibility, the staff chaplain will coordinate and use total UMT assets from both the TDA and TOE units. The installation chaplain assistant NCOIC advises the chaplain on chaplain assistant assignments, training, and other critical support issues.

SPECIALIZED RELIGIOUS SUPPORT

7-56. The Army is committed to supporting its families during routine operations and during deployment. Commanders employ the full range of installation resources to support families.

7-57. The provision of religious support to families is the responsibility of the installation chaplain. The Family Life UMT and the hospital UMT assist in their related areas in providing support. When not deployed, each unit's UMTs also have a responsibility to support their unit's families.

7-58. When units are deployed the RC will provide UMTs to assists in sustaining support to families. UMTs support the commander's Family Readiness Group (FRG) initiatives and also provide a vital role in supporting families when the unit's soldiers are deployed. Though not specifically a UMT responsibility, FRGs have a traditional relationship with the UMTs.

FAMILY LIFE UMT

7-59. The Family Life Chaplain's primary mission is to train UMTs and to direct the Chaplain Family Life Center (CFLC). Family life ministry consists of three major elements: enrichment, prevention, and intervention. These elements are expressed through educational programs, family wellness development programs, and marriage and family counseling.

7-60. The Family Life Chaplain consults with senior chaplains to determine the specific training needs for subordinate units. Each Family Life Chaplain tailors the training program to meet the needs of the installation and the mission of the tactical units. Training could be in the form of technical assistance, educational resources, and specific training for UMTs. In addition, the Family Life Chaplain helps UMTs prepare families for the stress of mobilization and deployment. This support is directly related to the installation's mission as the force projection platform. Each Family Life UMT supports the UMTs of the tactical units on the installation. They will tailor the training program to meet the needs of the installation and the mission of the tactical units on the installation. For example, a Family Life UMT will consult with a Division UMT to determine the division's specific training needs in family life issues. By training the Division's UMTs in family issues, the Family Life Chaplain serves as a direct link between the installation chaplain and the UMTs in sustaining the force. Training topics include—

- Ministry to FRGs.
- Family stress.
- Communication skills.
- Family counseling skills.
- Reunion after deployment.
- Grief and loss.

7-61. A healthy home environment directly influences the soldier's performance of duty. With its support of the family, the family life ministry program sustains the soldier's spiritual fitness in peacetime and during combat.

HOSPITAL UMT

7-62. The Army Medical Department preserves the fighting strength of the Army, maintains the health of soldiers, and supports family members with medical services.

7-63. Chaplains and chaplain assistants assigned to hospitals help patients and families deal with medical problems and crises. Hospital chaplains at all levels have special training and skills that equip them to do the following:

- Assist people with feelings of fear, loneliness, anger, failure, and loss of faith.
- Provide pastoral care for those suffering from critical or chronic illnesses or injuries.
- Prepare religious support contingency plans for mass casualty situations.
- Facilitate support groups for patients, their families, and staff.
- Serve as members of interdisciplinary case management teams and hospital committees.
- Provide pastoral care to hospital staff.
- Provide training for staff members and UMTs.

CONFINEMENT UMT

7-64. Most Army installations do not have local installation confinement facilities. Soldiers are usually held in confinement by special arrangement with local authorities. The installation staff chaplain provides pastoral care for locally confined soldiers in keeping with the guidance of the installation commander. The United States Disciplinary Barracks (USDB) is provided a UMT for RS. UMTs provide religious support for both prisoners and facility staff members. This religious support includes performing or providing for the following:

- Worship
- Religious education
- Counseling
- Moral leadership
- Spiritual fitness training
- Family support

GARRISON SUPPORT UNITS

7-65. Garrison Support Units (GSU's) come from the RC to support base operations (BASOPS) during continuing operations, and assist in other installation missions during Annual Training (AT) periods. Each GSU has UMT assets within their TDA. (See The Chief of Chaplains Crisis Management UMT Assets Handbook, October 2000.)

7-66. When installations are called upon to act as projection platforms, they require augmentation to meet their force projection responsibilities. The installation's increased religious support mission also requires additional support and augmentation. The scope of the installation's religious support mission encompasses support to soldiers remaining on the installation, soldiers processing through the installation; families of soldiers assigned to the installation, families of soldiers processing through the installation, and increased religious support to the installation hospital. The requirement for religious support augmentation will increase as the level of mobilization increases.

Training.

7-67. The supported Installation Chaplain will provide training to the GSU UMT personnel.

Organization

7-68. UMTs will be embedded in the GSU. The number of UMTs required varies from installation to installation. The number of UMTs is determined by applying an allocation rule to the number of projected personnel associated with the each installation's mobilization mission.

SUMMARY

7-69. The installation staff UMT is a critical part of the overall contingency plans of the commander. They provide a trained ready force of UMTs to meet the full spectrum operational mission of the U.S. Army. To adequately support the future requirements of an Army capable to conducting prompt and sustained operations, installation UMTs must plan and execute the religious support mission in support of the National Military Strategy, the OCCH (Office of the Chief of Chaplains) strategic plan, and the installation commander's CMRP. The installation staff UMT must—

- Support the UMTs assigned to war-fighting units.
- Focus on Army and Chaplaincy core competencies.
- Implement the command CMRP for readiness, mobilization, and deployment.
- Provide quality and effective training for UMTs.
- Maintain stewardship of religious support assets.
- Provide the projection base for force projection and quality ministry for families and tenant units.

RESOURCES

7-70. The following is a list of main resources utilized for writing and updating this doctrine.

- AR 165-1.
- JP 1-05.
- JP 5-00.2.
- JP 3-0 (100-5).
- JP 3-10.
- FM 16-1.
- FM 25-100.
- FM 25-101.
- FM 100-11.
- FM 100-9.
- FM 22-100.
- FM 100-20.

- FM 100-19.
- DA Pam 165-3.
- National Military Strategy
- Joint Vision 2010
- Army Vision 2010
- U.S. Army Chaplaincy Strategic Plan FY 2000-FY 2005
- Lessons Learned from past and current operation
- Lessons Learned from the training centers
- The Chief of Chaplain's UMT Crisis Assets Handbook, October 2000

Appendix A

RELIGIOUS SUPPORT IN CIVIL MILITARY OPERATIONS

Throughout U.S. history, chaplains have supported commanders through advisement on matters pertaining to the interface between Religious Support Operations and Civil Military Operations (CMO). This appendix provides the necessary guidelines for chaplains to engage in this role as required by their commanders. (See Appendix G for discussion on CD-C) UMTs are reminded that CMO support is a secondary responsibility and that the personal delivery of religious support is always the UMT's imperative.

ADVISING THE COMMANDER

A-1. Chaplains will support the commander through advisement in the following areas that may influence CMO:

- Indigenous religious, ethnic, and cultural influences.
- Beliefs, practices, and customs of religious groups in the AO.
- Religious issues related to displaced civilians.
- Human welfare needs as requested by religious NGOs and IOs.
- Relations with indigenous religious leaders when directed by the commander.

A-2. Under Title X of the U.S. Code, Chaplains should not perform the following:

- Direct participation in negotiations or mediations as sole participant.
- Human intelligence (HUMINT) collection and/or target acquisition.

PROVIDE AND PERFORM RELIGIOUS SUPPORT

A-3. Chaplains will provide worship and pastoral care in Civil Military Operations in accordance within the following guidelines:

- The priority of responsibility for Army Chaplains is to provide worship opportunities and pastoral care to soldiers and authorized civilians in the AO.
- Chaplains will provide for pastoral counseling, visitation, religious education, spiritual fitness training, moral leadership training, critical event debriefing and the provision of religious literature and items.
- Chaplains will identify and notify the command of meaningful opportunities for soldiers to participate in Civil Military Operations to promote the morale and spiritual well being of deployed soldiers.

- In coordination with the Civil Military Operations Center, chaplains may provide certain religious support to refugees and displaced persons when directed by the Commander, and after consultation with Staff Judge Advocate (SJA). In such cases, it is critical to avoid any activities that can be construed as proselytizing among refugees or displaced persons.
- Every effort to include indigenous religious leadership to meet these needs should be taken.
- *CMO and RS tasks will need to be coordinated to ensure success.*

CMO Tasks	Religious Support Tasks
Advise commander of legal obligations and moral considerations.	Advise commander of moral considerations as affected by religion.
	Advise command on accommodating religious practices in AO.
Establish policy for the handling of dislocated civilians.	Monitor care of displaced civilians to ensure it is moral/ethical.
Plan displaced civilian operations.	
Coordinate displaced civilian operations.	Advise command on religious support plan for displaced civilians.
Provide displaced civilian support.	
Plan humanitarian/civic assistance activities.	Plan religious support activities. Assist CMO personnel in coordinating with local religious leaders for implementation of CMO activities.
Provide public welfare support.	Assist command in meeting human welfare needs in conjunction with religious NGOs & IOs –clothing, food shelter.
Coordinate health, welfare, and morale services.	
Provide civilian supply support.	
Provide food and agriculture support.	
Provide information on cultural considerations.	Advise leadership on beliefs, practices, and customs of religious groups in AO
Provide cultural affairs support.	Advise CMO personnel on the role/influence of religion in cultural affairs.
	Assess distinct religious, cultural, ethnic, and gender ministry needs
	Brief the commander of religious influences on contingency operations
Provide language support.	
Provide arts, monuments and archives support.	Advise command on impact of local religions, sites and facilities which affect the mission
	Advise command on the training of soldiers with respect to religious beliefs, practices, sites and facilities

CMO Tasks	Religious Support Tasks
Provide legal support.	
	Implement religious support in a Combined and/or Joint Area of Operation with Consideration to Host Nation Religious Background.
Prepare an area assessment.	Performs analysis of local religion as it affects mission.
Prepare an area study.	Prepares the RS annex to the OPORD, which includes impact of indigenous religions on AO and mission.
Implement the command map program.	Provide input to CMO personnel regarding local religions.
CMO must analyze the effect of FNS on the morale of U.S. soldiers and on the psychological condition of the local populace.	Advise leadership on negative changes in solder attitude on local population.
Plan disaster relief operations.	Provide religious support for humanitarian assistance survey team (HAST) operations.
Plan noncombatant evacuation operations.	
Commanders must segregate civilians from EPWs and civilian internees to protect them as required by international law.	Coordinate religious support for Enemy Prisoners of War.
Coordinate operations.	Coordinate religious support activities with CMO personnel and local religious leaders.
Perform command functions.	Provide religious support to CMO personnel.

Appendix B

COMBAT AND OPERATIONAL STRESS CONTROL AND RELIGIOUS SUPPORT

References: FM 8-51.
FM 22-51.
Textbook of Military Medicine: War Psychiatry, Part I; Warfare, Weaponry, and the Casualty Office of the Surgeon General of the United States Army, Falls Church, Virginia: 1995

INTRODUCTION

This appendix addresses the general role of the Unit Ministry Team in the commander's program for Combat and Operational Stress Control. Unit Ministry Teams, imbedded within units down to battalion level, provide immediate support to leaders in fulfilling their battle fatigue responsibilities. Unit Ministry Teams also assist in training leaders to recognize battle fatigue identification and intervention responsibilities. In cooperation with unit medical personnel, Unit Ministry Teams serve as a primary referral agency to mental health resources.

BACKGROUND

B-1. The negative impact of Combat and Operational Stress has been known by many names throughout the history of warfare: Nostalgia, Soldier's Heart, Shell Shock, War Neurosis, Combat Stress Reaction, and Battle Fatigue. Combat Stress Behaviors cover the full range of behavior in combat from the highly positive to the completely negative.

B-2. Positive Combat Stress Behaviors include heightened alertness, strength, endurance, and tolerance of discomfort. Examples of positive Combat Stress Behaviors would include the strong personal trust, loyalty, and cohesiveness, which develops among, peers in small units, as well as between leaders and subordinates.

B-3. Dysfunctional Combat Stress Behaviors come in two forms: Misconduct Stress Behaviors and Criminal Acts, and Battle Fatigue.

B-4. Misconduct Stress Behaviors and Criminal Acts range from simple UCMJ infractions to outright criminal acts such as killing enemy prisoners of war, threatening or killing superiors, or looting, pillaging, and rape.

B-5. Battle Fatigue, also referred to as Combat Stress Reaction or Combat Fatigue, is the distress and impaired performance resulting from the accumulated stressors of the combat or operational situation. Signs of Battle Fatigue would include hyper alertness; fear and anxiety; grief, self-doubt, and guilt; loss of confidence and hope; impaired duty performance; erratic actions and outbursts; freezing and immobility; terror, panic, and flight; and impaired speech, vision, touch, and hearing. In extreme cases Battle Fatigue may result in paralysis and hallucinations.

THE BASICS OF TREATMENT

B-6. The principles for stabilizing treatment of soldiers experiencing Battle Fatigue are summarized by the acronym **PIES**.

- **Proximity**: Battle Fatigue cases are treated as far forward as tactically possible and in close proximity with the soldier's unit with the goal of returning the soldier to duty after a period of several days.
- **Immediacy**: Immediate initiation of treatment.
- **Expectancy**: Expectation of rapid and full recovery and return to duty.
- **Simplicity**: Simplicity of approach, according to the four Rs:
  - *Reassurance*: Reassurance of normality.
  - *Rest*
  - *Replenishment*: Replenishment of nutrition, hydration, hygiene and sense of physical well being.
  - *Restoration*: Restoration of confidence through talk and activities.

B-7. It is important to note that the soldier's identity as a soldier is maintained and reinforced. Throughout treatment Battle Dress Uniforms are worn, not hospital pajamas. Rank distinctions and military courtesy are maintained, and these soldiers are responsible for self-care and helping others.

RELIGIOUS SUPPORT AND COMBAT/OPERATIONAL STRESS CONTROL

B-8. Soldiers' inner resources are generally rooted in their religious and spiritual values. In combat, soldiers often show more interest in their religious beliefs. When religious and spiritual values are challenged by the chaos of combat, soldiers may lose connection with the inner resources that have sustained them. Then they become targets of fear, and hopelessness, and eventually fall victim to Battle Fatigue. Further, they become at risk for Misconduct Stress Behaviors. The Unit Ministry Team is the primary resource available to soldiers experiencing these dilemmas and seeking to refocus their spiritual values.

The Unit Ministry Team's Role

B-9. Unit Ministry Teams provide preventative, immediate, and restorative spiritual and emotional support and care to soldiers experiencing Battle Fatigue.

B-10. Preventative Religious Support. The Unit Ministry Team assists in preventing Battle Fatigue and Misconduct Stress Disorders through Spiritual Fitness training. Ministry of Presence with soldiers, assigned DA civilian and contractor personnel is critical. The Unit Ministry Team provides a stabilizing influence on personnel, and assists them in strengthening and regaining personally held spiritual values. Preventative activities include—

- Worship opportunities.
- Private and group prayer opportunities.

- Religious literature and materials.
- Scripture readings.
- Sacraments and ordinances.
- Opportunities for working through frustration, fear, anxiety, and anger.
- Assistance to personnel and families prior to deployment, emphasizing family strengths.

B-11. Immediate Religious Support. The Unit Ministry Team assists commanders in the identification of personnel experiencing negative reactions to combat and operation stress: Battle Fatigue and Misconduct Stress Behaviors. The Unit Ministry Team works closely with the unit's leaders and medical personnel to care for Battle Fatigue cases through religious support and comfort. Immediate Religious Support activities may include—

- Presence with the soldier.
- Conversation focused upon fears, hopes, and other feelings.
- Conversation focused upon forgiveness.
- Prayer with personnel.
- Prayer for fallen comrades and Memorial Ceremonies and Services.
- Rites, sacraments, and ordinances, as appropriate.
- Sacred Scripture.

B-12. Restorative Religious Support. Following an operation, a unit may require reconstitution. Surviving soldiers may need to rebuild emotional, psychological, and spiritual strength. Depending upon the spiritual, emotional, and physical condition of the unit's soldiers, the organic Unit Ministry Team may need augmentation from higher echelons or other units. Restorative Religious Support activities may include—

- Worship, sacraments, rites, and ordinances.
- Memorial ceremonies and services.
- Religious literature and materials.
- Grief facilitation and counseling.
- Reinforcement of the soldiers' faith and hope.
- Opportunities for soldiers to talk about combat experiences and to integrate those experiences into their lives.

COMBAT STRESS BEHAVIORS

POSITIVE COMBAT STRESS BEHAVIORS

DYSFUNCTIONAL COMBAT STRESS BEHAVIORS

MISCONDUCT STRESS BEHAVIORS AND CRIMINAL ACTS

BATTLE FATIGUE

POSITIVE COMBAT STRESS BEHAVIORS	MISCONDUCT STRESS BEHAVIORS AND CRIMINAL ACTS	BATTLE FATIGUE
UNIT COHESION LOYALTY TO BUDDIES LOYALTY TO LEADERS IDENTIFICATION WITH UNIT TRADITION SENSE OF ELITENESS SENSE OF MISSION ALERTNESS AND VIGILANCE EXCEPTIONAL STRENGTH AND ENDURANCE INCREASED TOLERANCE TO HARDSHIP DISCOMFORT PAIN INJURY SENSE OF PURPOSE INCREASED FAITH HEROIC ACTS COURAGE SELF-SACRIFICE	MUTILATING ENEMY DEAD NOT TAKING PRISONERS KILLING ENEMY PRISONERS KILLING NONCOMBATANTS TORTURE, BRUTALITY KILLING ANIMALS FIGHTING WITH ALLIES ALCOHOL AND DRUG ABUSE RECKLESSNESS, INDISCIPLINE LOOTING, PILLAGE, RAPE FRATERNIZATION EXCESSIVELY ON SICK CALL NEGIGENT DISEASE, INJURY SHIRKING, MALINGERING COMBAT REFUSAL SELF0INFLICTED WOUNDS THREATENING/KILLING OWN LEADERS ("FRAGGING") ABSENCE WITHOUT LEAVE DISERTION	HYPERALERTNESS FEAR, ANXIETY IRRITABILITY, ANGER, RAGE GRIEF, SELF-DOUBT, GUILT PHYSICAL STRESS COMPLAINTS INATTENTION, CARELESSNESS LOSS OF CONFIDENCE LOSS OF HOPE AND FAITH DEPRESSION, INSOMNIA IMPAIRED DUTY PERFORMANCE ERRATIC ACTIONS, OUTBURSTS FREEZING, IMMOBILITY TERROR, PANIC RUNNING TOTAL EXHAUSTION APATHY LOSS OF SPEECH SKILLS AND HEARING WEAKNESS, PARALYSIS HALLUCINATIONS DELUSIONS

Appendix C

CRISIS INTERVENTION MANAGEMENT AND CRITICAL EVENT DEBRIEFINGS

BACKGROUND

There are currently three major schools of Crisis Intervention Management, of which Critical Event Debriefing (CED) by a variety of names, are a part. These are the International Critical Incident Stress Foundation's Critical Incident Stress Management (CISM), the National Organization for Victim Assistance (NOVA) Group Crisis Intervention (GCI), and the American Red Cross Disaster Response. In each case, CED is a component of the larger Crisis Intervention model, and training and certification requirements vary from model to model.

PURPOSE

C-1. Crisis Intervention helps small units soon after exceptionally traumatic events to—

- Quickly restore unit cohesion and effectiveness.
- Reduce short-term emotional and physical distress.
- Mitigate long-term distress and "burnout".
- Facilitate the transition from traumatic event (s) to a sense of normalcy and aids in future transitions.

REQUIREMENT

C-2. Leaders of small units should ordinarily conduct an after action review (AAR) when a mission is completed. After an exceptionally distressing mission or event, when METT-TC dependent, unit leaders may coordinate crisis intervention. Members of the Unit Ministry Team are equipped to provide this intervention.

C-3. Situations that may warrant crisis intervention include—

- Death of the unit member.
- Death or suffering of noncombatants (especially women and children).
- Handling of human remains, management of carnage, extensive disaster devastation.
- Friendly fire incident.
- Situation involving serious error, injustice, or atrocity.

C-4. Crisis intervention may be conducted by the Unit Ministry Team or may be coordinated with Mental Health/Combat Stress Control or other medical personnel.

PERSPECTIVE

C-5. Participants in crisis intervention are normal people who have survived an abnormal situation. CED is a generic term for a specific tool within the broader practice of crisis intervention management. CED is neither therapy nor clinical counseling. Rather, it is basic and wise interventive maintenance for the human spirit.

THE CRITICAL EVENT DEBRIEFING

C-6. WHO

- A functional organization: crew, team, squad, or platoon.
- Strangers, civilians, family members thrown together into a traumatic situation by chance.
- Normally includes only those directly involved in the event.
- Chain of Command included only if involved in the event.
- Includes members of identified Crisis Intervention Team.
- Media and outsiders excluded.

C-7. WHAT

Normally, the team consists of a debriefing leader and an assistant, with one additional assistant for every 10 people in a group of more than 20.

C-8. WHEN

- Most effective when conducted 12-72 hours after the event.
- Allow for enough rest and recovery for all to be alert and involved.
- After completing an operation with no expected intense follow-on operations for next 72 hours.
- Expected duration: 2-3 hours.
  When ongoing tactical operations prohibit CED, a briefer "defusing debriefing" focusing on cognitive reactions to the situation can be substituted

C-9. WHERE

- An emotionally neutral place, relatively safe from enemy action, distraction, observation.
- Reserve position or assembly area.
- With enough light to see all participants.
- Shelter from bad weather.

C-10. HOW

- **Introductory Phase**
  Purpose: To introduce the Crisis Intervention Team and explain the process.
  Chaplain or Leader, explains the grounds rules:
  - Personal information or feelings shared in the group during CED are not to be shared outside the group

---

Note: This does not override legal or moral responsibility to report violations of UCMJ or the Law of Land Warfare.

  - No notes or recordings.

  - No breaks are scheduled, but anyone may leave as needed, returning as soon as possible.
  - No one is expected to speak after the Fact Phase.
  - Each participant speaks for self, not others.
  - Everyone is equal during CED. All ranks speak frankly, with proper courtesy.
  - CED is not an AAR, but a discussion to clarify what happened and to restore sense of well-being.
  - Fact-finding, not Fault finding.
  - "Facts" include personal reactions the event.
  - The Team is available following the CED.
- **Fact Phase**
  Purpose: Reconstruct the event in detail, in chronological order, as an unbroken "historical time line" viewed from all sides and perspectives. Chaplain/Leader encourages all participants to start their stories before the event occurred and to work up to it.
  Participants—
  - The first person involved in the event is asked to tell his version of the story- how it started, his role (duty position), and what he saw, heard, smelled, and did-step by step.
  - One by one, participants are drawn into the first person's story and then asked to describe their observations and actions.
  - The Chaplain/Leader encourages participation. Everyone is asked to speak.
  - When disagreements over what occurred arise, the Chaplain/leader elicits group observations to clarify memories.
- Thought Phase
  Purpose: To personalize the event.
  Chaplain/Leader facilitates transition from facts to the personal.
  Participants are asked to share—
  - "What were you thinking as the event started?"
  - "What went through your mind when you saw, smelled, or did...?"
  - "What was your first thought when you came off 'automatic' or 'autopilot'?"
- **Reaction Phase**
  Purpose: Identify and ventilates feelings (emotions) raised by the event. Chaplain/Leader emphasizes that all emotional reactions deserve to be expressed, respected, and listened to.
  Participants are encouraged to share "reactions":
  - "What was the worst thing about the event?"
  - "How did you react when it happened?"
  - "How are you feeling about that now?"
  - "If you could change one thing about the event, what would it be?"

  Chaplain/Leader listens for common themes, feelings, and misperceptions—
  - Feelings of anger at others for not helping.
  - Blaming self or others for things beyond one's control.

- Feeling changed, different, worse than everyone else, cut off from others.

Participants are asked to describe physical reactions to the event. Common symptoms include—

- Gastro-intestinal distress
- Frequent urination
- Loss of bowel and bladder control
- Loss of sexual interest
- Heart pounding
- Shortness of breath
- Muscle, back, neck, and head ache
- Trembling, jumpiness, and startle reactions

- **Reframe Phase**

  Purpose: To transition from emotional reaction to the cognitive. Chaplain/Leader facilitates the search for meaning in the event. Participants are encouraged to derive personal meaning in the event.

  - "What lessons could be learned from this event?"
  - "What is something positive that you will take away from this experience?"
  - "What good can be found in this tragic situation?
  - "What are you proud of in this event?"

- **Teaching Phase**

  Purpose: To reassure by teaching the participants that feelings and stress symptoms are normal reactions to abnormal conditions. Symptoms may last a while, but can be expected to resolve in time. If they don't resolve in time, seeking professional help is advised.
  Chaplain/Leader summarizes the thoughts, feeling, and symptoms expressed by the group, reemphasize normality; and reduce feelings of uniqueness, weakness, or injury.
  No predictions or glamorization of long-term disability should be given.
  The Unit Ministry Team may schedule additional training in stress management, coping strategies, grief process, and anger management at later dates.

- **Reentry Phase**

  Purpose: Complete and close the CED.
  Chaplain/Leader gives final invitation for comments and makes a summary statement. Follow-up resources are discussed.
  It is important for Unit Ministry Team personnel to make themselves available for follow-on conversations with individual participants following the CED.

Appendix D

INSTRUCTIONS FOR USING THE MODEL RELIGIOUS SUPPORT MATRIX

GENERAL INFORMATION (ADAPTED FROM FM 101-5)

This appendix provides a sample Religious Support Matrix. The matrix provides a highly visible, clear method to ensure all religious support providers are synchronized. It shows the relationships between activities, units, support functions and key events. Although the unit order will not appear in matrix format, the staff can write their annex as an execution matrix. The religious support appendix as part of the Service Support annex can be done in matrix format. An execution matrix depicts when and where specific supporting actions must occur.

HEADER INFORMATION

D-1. *Classification*. Army Regulation 380-5 contains the detailed description of classification procedures. If the entire plan is unclassified, no classification is required on the appendix.

D-2. *Annexes, Appendixes, Tabs, and Enclosures*. FM 101-5, 31 May 1997, p. H-7ff gives details on how annexes, appendixes, tabs, and enclosures are used. Follow unit SOP and include a matrix as shown in the sample in Figure D-1. The matrix should be included as an appendix to the Service Support Annex. An Arabic number designates the appendix; a capital letter designates the Annex . The Service Support Annex should include the Religious Support Matrix in the list of appendixes along with the other appendixes, such as Personnel and Legal.

D-3. *References*. References list information related to the matrix. The information should help the intended users understand the matrix. Map sheet series numbers may be included. The unit SOP does not need to be referenced. (Cf. FM 101-5, 31 May 1997 p. H-14)

D-4. *Time Zone*. The time zone used throughout the order is placed here. Operations across several zones use ZULU time. If the local time is chosen, the appropriate time zone suffix is used. The time zone is not based on geographical location but rather on the relationship to ZULU time. For example, Central Standard Time would normally be SIERRA. However, during daylight savings time ROMEO would be the appropriate suffix.

D-5. *DTG*. Date time group (DTG) refers to the time of signature in day, hour with suffix, month, and year format. E.g. 260900S JAN04 would be 26 January 2004 at 0900 hr in the SIERRA time zone.

D-6. *Task Organization*. Task Organization is found in Annex A of the OPORD/OPLAN. The UMT coverage plan is made by matching the UMTs with the units listed in the Task Organization. The unit designator like "Iron Horse" goes under the Unit name e.g. (2BCT). List the chaplain and assistant who provide the coverage. When the units are Task Force organized the chaplain of the Task Force is automatically responsible for all the units that would show up in their level Task Organization. If the coordinated Religious Support Plan calls for chaplains to cover units that are not a part of their unit's Task Organization for the entire operation, note those exceptions under the units listed to the right. If the exceptions change by phase, list the coverage in the concept of operation under units. List the chaplain and assistant who provide coverage and indicate the denomination or faith group of chaplains to more completely track coverage needs in the unit AO.

- Enemy Situation. Normally a reference to the Intelligence Annex will suffice. This presupposes that the chaplain has coordinated with the S2/G2 during the Religious Area Analysis (RAA) phase of planning. If there is anything specific about the enemy related to the RS mission that is not listed in the Intelligence annex, then it would go here.
- Friendly Situation. Normally, the basic order can be referenced here. Significant religious issues for the friendly forces are also added here if necessary. Note Religious Preference Profiles (RPPs) that differ from the norm. National Guard units can at times take local flavors. For instance, a National Guard unit in Utah may have a high percentage of soldiers who are members of the Church of Jesus Christ of Latter Day Saints. The Hawaii National Guard might have a unique local flavor as well. Anything that affects your religious support mission that is not the norm is noted here.
- Mission. State the Religious Support tasks and their purpose. Mission is to ____________ in order to ______________.
- Concept of Operations.
  - ❑ Location. Location answers the question, "Where will people go for services." During the different phases as the senior chaplain changes location people may have to go different places in order to receive or coordinate for Religious Support Services. Grid designators, Area of Operation, ISB chapel, are types of information that you can use.
  - ❑ Units. Note here exceptions to the coverage plan from what is listed under Task Organization. The purpose of the block is to state what UMT is providing which unit service. If the aviation chaplain cover the smoke platoon in the first phase but the field artillery chaplain covers the smoke platoon in the second phase, then list it here.
  - ❑ Time Frame. The time that is referred to here is the time schedule for services rendered. This is not the beginning and the ending of the phase. The start and ending of a phase can be listed under the Title of the Phase. Are there any special services or special religious holidays that happen during the operation? These can be listed here. At battalion level you may be able to state when you are going to hold a service for a particular company.

- UMT Priority of Support. UMT Priority of support refers to the services rendered by the UMT. While the unit is in the ISB the priority might be generally nurture the living. Then you would want to spell out just what that means: religious services, counseling opportunities, classes on stress, etc. During the battle the priority might be caring for casualties in general. Then what does that mean specifically? You could list praying for casualties, ministering to care givers, etc. During the post-battle phase honoring the dead might be the priority. During redeployment nurturing the living might include reunion briefings.
- Priority of Support. Priority of Support refers to the units that will receive the services offered by the UMT. While in the ISB, you may give priority to the scouts because of the danger of their mission and because they will be hard to reach later. Often, the religious support priority matches the commander's priority of support as found in the order. Normally, the main effort receives the priority of support.

D-7. *Coordinating Instructions.* Coordinating Instructions tells the unit how they will coordinate for services. How do people coordinate for religious services? How does the unit coordinate for Roman Catholic coverage or some other minority faith need? Before chaplains go on a civil affairs type of mission what coordination needs to take place. How are extraordinary Ministers of the Eucharist being used? How will denominational service leaders (DSLs) function or will they?

D-8. *Service Support.* State how many days of supply each UMT will need. State how re-supply is conducted. Normally, re-supply is through S4 channels.

D-9. *Command.* Report times and formats can be listed here. You can list the specific chain of succession for this operation. You should have the general procedures for the chain of succession in your TSOP.

D-10. *Signal.* List primary and alternate means of communication. List specific information that is needed to contact the chaplain who is responsible for the appendix. If it is a Brigade appendix it is the brigade chaplain. If it is a battalion appendix, then it should be the battalion chaplain.

D-11. *Safety.* List specific safety concerns. Could state that the UMTs need to do a Risk Assessment before certain types of missions (e.g. Civil Affairs or movement to new locations).

D-12. Extraneous Information.

- Name and position example Dolinger, TF CH
- Time Zone: The time zone is the same one used throughout the order.
- OFFICIAL: If the commander signs the original this is not used. The signed copy is a historical record and should remain in headquarters files. The preparing officer signs under OFFICIAL when the commander does not sign it.
- Types of services available
- Designation and location of unit or place providing the service.
- Schedules for services

- Specific missions for units (Special religious groups or holidays)
- Concept of operations.
  - Priority of effort
  - Location
  - Procedures
  - Policies
  - Requirements
  - Phases are specific parts of an operation that differ from those that precede or follow. Phases are indicated by time, distance, terrain, or event occurrences. These will be identified in the order.

CLASSIFICATION

APPENDIX 4 (RELIGIOUS SUPPORT MATRIX) TO ANNEX I (SERVICE SUPPORT) to OPLAN 10-01 OPERATION OVERLORD

REF: AR 165-1, JP 1-05, FM 1-05, RB 16-100	TIME ZONE: ZULU	DTG: 060300JUN44

TASK ORGANIZATION: See Annex A (Task Org) for each phase	**Unit Designator** CH Goode Assistants SSG Nock	**Subordinate Unit Designator** CH Poling Assistants SSG Bizarri	**Subordinate Unit Designator** CH Fox Assistants SGT Woods	**Subordinate Unit Designator** CH Washington Assistants SPC Duff	**Subordinate Unit Designator** CH Watters Assistants SGT Titus	**Subordinate Unit Designator** CH Rayburn Assistants SGT Calvin	**Subordinate Unit Designator** CH Spurgeon Assistants SGT Luther	**Subordinate Unit Designator** CH Augstine Assistants PVT Delta

1. SITUATION	**1a. ENEMY SITUATION:** Annex B (Intelligence)	**1b. FRIENDLY SITUATION:** Basic Order

2. MISSION: Provide RS to dispersed elements of BCT NLT 081100JUN44 and provide RS to MASCAL sites as directed.

3. EXECUTION: *(intent)*

3a. Concept of Operations	Phase I Title of Phase	Phase II Title of Phase	Phase III Title of Phases	Phase IV Title of Phase
Location	UMTs cover units starting w/ those closest to the front	UMTs locate at Casualty Collection Points	UMTs cover CCPs and units first to attack	TBD
Units	EN UMT covers smoke platoon	FA UMT covers smoke platoon	TBD	TBD
Time Frame	Average 5 services daily per UMT	Average 1 service daily plus personal casualty ministry	Average 2 services daily	TBD
UMT Priority of Effort	Religious services, Prayer, Counseling	Prayer for casualties and dying, ministry to care givers	Prayer for casualties, religious services	TBD
Priority of Support	IN TF, EN BN	IN TF, AR TF	EN BN, FA BN	TBD

3b. COORDINATING INSTRUCTIONS.

(1) All UMTs going on CMO missions must coordinate w/ the BDE UMT.

(2) Units desiring Roman Catholic Coverage must coordinate w/ the BDE UMT.

(3) UMTs will report service times IAW SOP.

4. SERVICE SUPPORT

(1) UMTs will carry 30 days of supply.

(2) Resupply is through unit S4 channels.

5. COMMAND AND SIGNAL

5a. COMMAND.

(1) Chain of succession is FSB, EN, FA UMTs

(2) BDE UMT will use TOC as primary location.

(3) Send in reports IAW TACSOP

5b. SIGNAL.

(1) Primary means of communication is FBCB2, DVNT, VIASAT, BDE Admin/log FM net, Messanger

(2) Radio suffix for all Chaplains is 9, expander for Chaplain Assistants is 0.

5c. SAFETY.

UMTs will conduct Risk Assessment prior to all convoy movement and CMO missions.

ACKNOWLEDGE:

OFFICIAL: MOSES, BDE Chaplain

NAME (HYATT)

Rank (COL)

Figure D-1. Sample Model Religious Support Matrix

Appendix E

UNITED STATES CODES RELEVANT TO RELIGIOUS SUPPORT OPERATIONS

This appendix contains extracts from the USC relevant to RS. The extracts range from general military law to laws governing conscientious objectors, accommodation of religious practices and Title 18, Posse Comitatus Act.

E-1. Section 3073. Chaplains. There are chaplains in the Army. The Chaplains include—

- The Chief of Chaplains.
- Commissioned officers of the Regular Army appointed as chaplains.
- Other officers of the Army appointed as chaplains in the Army.

E-2. Section 3547. Duties: chaplains; assistance required of commanding officers—

- Each chaplain shall, when practicable, hold appropriate religious services at least once on each Sunday for the command to which he is assigned, and shall perform appropriate religious burial services for members of the Army who die while in that command.
- Each commanding officer shall furnish facilities, including necessary transportation, to any chaplain assigned to his command, to assist the chaplain in performing his duties.

E-3. Section 3581. Command: chaplains. A chaplain has rank without command.

E-4. Section 774. Religious apparel: wearing while in uniform—

- General Rule. - Except as provided under subsection (b), a member of the armed forces may wear an item of religious apparel while wearing the uniform of the member's armed force.
- Exceptions. - The Secretary concerned may prohibit the wearing of an item of religious apparel.
  - In circumstances with respect to which the Secretary determines that the wearing of the item would interfere with the performance of the member's military duties; or
  - If the Secretary determines, under regulations under subsection (c), that the item of apparel is not neat and conservative.

- Regulations. The Secretary concerned shall prescribe regulations concerning the wearing of religious apparel by members of the armed forces under the Secretary's jurisdiction while the members are wearing the uniform. Such regulations shall be consistent with subsections (a) and (b).
- Religious Apparel Defined. In this section, the term "religious apparel" means apparel the wearing of which is part of the observance of the religious faith practiced by the member.

E-5. Section 6031. Chaplains: divine services—

- An officer in the Chaplain Corps may conduct public worship according to the manner and forms of the church of which he is a member.
- The commanders of vessels and naval activities to which chaplains are attached shall cause divine service to be performed on Sunday, whenever the weather and other circumstances allow it to be done; and it is earnestly recommended to all officers, seamen, and others in the naval service diligently to attend at every performance of the worship of Almighty God.
- All persons in the Navy and in the Marine Corps are enjoined to behave themselves in a reverent and becoming manner during divine service.

E-6. Section 8547. Duties: chaplains; assistance required of commanding officers—

- Each chaplain shall, when practicable, hold appropriate religious services at least once on each Sunday for the command to which he is assigned, and shall perform appropriate religious burial services for members of the Air Force who die while in that command.
- Each commanding officer shall furnish facilities, including necessary transportation, to any chaplain assigned to his command, to assist the chaplain in performing his duties.

DEPARTMENT OF DEFENSE DIRECTIVES

DODD 1300.6 – Conscientious Objectors

E-7. References:

- DOD Directive 1300.6, subject as above, May 10, 1968 (hereby canceled)
- DOD Directive 1332.14, "Administrative Discharges, " December 20, 1965
- Section 6(j) of the Universal Military Training and Service Act, as amended (50 U.S.C. App. 456(j))
- Section 3103, title 38, United States Code

Definitions

E-8. **3.1. Conscientious Objective 1-0**. A firm, fixed and sincere objection to participation in war in any form or the bearing of arms, by reason of religious training

E-9. **3.1.1. Class1-O Conscientious Objector**. A member who, by reason of conscientious objection, sincerely objects to participation of any kind in war in any form.

E-10. **3.1.2. 1-A-O Conscientious Objector**. A member who, by reason of conscientious objection, sincerely objects to participation as a combatant in war in any form, but whose convictions are such as to permit a Military Service in a non-combatant status.

E-11. **3.3.** Non-combatant service or non-combatant duties (1-A-O) used interchangeably herein)

E-12. **3.3.1.** Service in any unit of the Armed Forces which is unarmed at all times.

E-13. **3.3.2.** Service in the medical department of any of type Armed Forces, wherever performed.

E-14. **3.3.3.** Any other assignment the primary function of which does not require the use of arms in com bat provided that such other assignment is acceptable to the individual concerned and does, not require him to bear arms or to be trained in their use.

E-15. **3.3.4.** Service aboard an armed ship or aircraft or in a combat zone shall not be considered to be combatant duty unless the individual concerned is personally and directly involved in the operation of weapons

E-16. **3.4. Non-combatant Training**. Any training which is not concerned with the study, use or handling of arms or weapons.

E-17. **5.3.1.** In order to find that an applicant's moral and ethical beliefs are against participation in war in any form and are held with the strength of traditional religious convictions, the applicant must show that these moral and ethical convictions, once acquired, have directed his life in the way traditional religious convictions of equal strength, depth and duration have directed the lives of those whose beliefs are clearly found in traditional religious convictions. In other words, the beliefs are clearly found in traditional religious convictions. In other words, the beliefs upon which conscientious objection is based must be the primary controlling force in the applicant's life.

E-18. **5.3.2.** A primary factor to be considered is the sincerity with which the belief is held. Great care must be exercised in seeking to determine whether asserted belief is honestly and genuinely held. Sincerity is determined by an impartial evaluation of the applicant's thinking and living in its totality, past and present. Care must be exercised in determining the integrity of belief and the consistency of application. Information presented by the claimant should be sufficient to convince that the claimant's personal history reveals views and actions strong enough to demonstrate that expediency or avoidance of military service is not the basis of his claim.

E-19. **5.3.2.1.** Therefore, in evaluating applications the conduct of applicants, in particular their outward manifestation of the beliefs asserted, will be carefully examined and given substantial weight.

E-20. **5.3.2.2.** Relevant factors that should be considered in determining an applicant's claim of conscientious objection in the home and church; general demeanor and pattern of conduct; participation in religious activities; whether ethical or moral convictions were gained through training, study, contemplation, or other activity comparable in rigor and dedication to the processes by which traditional religious convictions are formulated; credibility of the applicant; and credibility of persons supporting the claim.

E-21. **5.3.3.1.** Church membership or adherence to particular theological tenets are not required to warrant separation or assignment to non-combatant training and service for conscientious objectors.

E-22. **5.3.3.2.** Mere affiliation with church or other group which advocates conscientious objection as a tenet of its creed is not necessarily determinative of an applicant's position or belief.

E-23. **5.3.3.3.** Conversely, affiliation with a church or group which does not teach conscientious objection does not necessarily rule out adherence to conscientious objection beliefs in any given case.

E-24. **5.3.3.4.** Where an applicant is or has been a member of a church, religious organization, or religious sect, and where his claim of conscientious objection is related to such membership, inquiry may properly be made as to the fact of membership, and the teaching of the church, religious organization, or religious sect, as well as the applicant's religious activity. However, the fact that the applicant's may disagree with, or not subscribe to, some of the tenets of his church does not necessarily discredit his claim. The personal convictions of each individual will be controlling so long as they derive from moral, ethical or religious beliefs.

E-25. **5.4.** The burden of establishing a claim of conscientious objection as grounds for separation or assignment to non- combatant training and service is on the applicant. To this end, he must establish clear and convincing evidence (1) that the nature or basis of his claim comes within the definition of and criteria prescribed herein for conscientious objection, and (2) that his belief in connection therewith is honest, sincere and deeply held. The claimant has the burden of determining and setting forth the exact nature of his request, i.e., whether for separation based on conscientious objection (1-O), or, for assignment to non-combatant training and service based on conscientious objection (1-A-O).

E-26. **6.3.** The applicant shall be personally interviewed by a chaplain who shall submit a written opinion as to the nature and basis of the applicant's claim, and as to the applicant's sincerity and depth of conviction. The chaplain's report shall include the reasons for his conclusions. This opinion and report will become part of the "case file." If the applicant refuses to participate or is uncooperative or unresponsive in the course of the interviews, this fact will be included in the statement and report filed by the chaplain and psychiatrist or medical officer.

E-27. **6.4.3.6.** The investigating officer's report, along with the individual's application, all interviews with chaplains or doctors, evidence received as a result of the investigating officer's hearing, and any other items submitted by the applicant in support of his case will constitute the record. The investigating officer's conclusions and recommended dispositions will be based on the entire record and not merely on the evidence produced at the hearings. A copy of the record will be furnished to the applicant at the time it is forwarded to the commander who appointed the investigating officer, and the applicant will be informed that he has the right to submit a rebuttal to the report within the time prescribed by the Military Service concerned.

E-28. **6.6.** The Secretary of a Military Service may delegate authority to approve applications to the commander exercising general court-martial jurisdiction (or equivalent level command for Reserve organizations) over the applicant. The completed record of a case approved in the field will be forwarded to the Headquarters of the Military Service concerned for appropriate disposition.

E-29. **6.9.** To the extent practicable under the circumstances, during the period applications are being processed and until a decision is made, every effort will be made to assign applicants to duties, which will conflict as little as possible with their asserted beliefs. Unless the Military Service concerned provides otherwise, an applicant shall be required to comply with active duty or transfer orders in effect at the time of his application or subsequently issued and received. During the period applications are being processed, applicants will be expected to conform to the normal requirement of military service and to perform such duties as are assigned. Applicants may be disciplined for violations of the Uniform Code of Military Justice while awaiting action on their applications.

DODD 1300.17 – Accommodation of Religious Practices within the Military Services

E-30. References:

- DOD Directives 1300.17, subject as above, June 18, 1985 (hereby canceled)\
- Public Law 98-525, section 554(d), DOD Authorization Act, 1985, October 19, 1984
- Public Law 100-180, section 508, DOD Authorization Act, 1988-89, December 4, 1987
- Title 10, United States Code, Chapter 47, Uniform Code of Military Justice

E-31. **3.1.** A basic principle of our nation is free exercise of religion. The Department of Defense places a high value on the rights of members of the Armed Forces to observe the tenets of their respective religions. It is DOD policy that requests for accommodation of religious practices should be approved by commanders when accommodation will not have an adverse impact on military readiness, unit cohesion, standards, or discipline.

E-32. **3.2.1.** Worship services, holy days, and Sabbath observances should be accommodated, except when precluded by military necessity.

E-33. **3.2.2.** The Military Department should include religious belief as one factor for consideration when granting separate rations, and permit commanders to authorize individuals to provide their own supplemental food rations in a field or "at sea" environment to accommodate their religious beliefs.

E-34. **3.2.3.** The Military Department should consider religious beliefs as a factor for waiver of immunizations, subject to medical risks to the unit and military.

E-35. **3.2.6.** Religious items or articles not visible or otherwise apparent may be worn with the uniform, provided they shall not interfere with the performance of the member's military duties, as discussed in subparagraph

3.2.7.5. below, or interfere with the proper wearing of any authorized article of the uniform.

E-36. **3.2.7.** Under Public Law 100-180, section 508 (reference (C)), members of the Armed Forces may wear visible items of religious apparel while in uniform, except under circumstances in which an item is not neat and conservative or its wearing shall interfere with the performance of the member's military duties.

E-37. **3.2.7.1.** Under this Directive, "religious apparel" is defined as articles of clothing worn as part of the doctrinal or traditional observance of the religious faith practiced by the member. Hair and grooming practices required or observed by religious groups are not included within the meaning of religious apparel, Jewelry bearing religious inscriptions or otherwise indicating religious affiliation or belief is subject to existing Service uniform regulation just as jewelry that is not of a religious nature.

E-38. **3.2.7.2.** In the context of the wearing of a military uniform, "neat and conservative" items of religious apparel are those that:

E-39. **3.2.7.2.1.** Are discreet, tidy, and not dissonant or showy in style, size, design brightness, or color.

E-40. **3.2.7.2.2.** Do not replace or interfere with the proper wearing of any authorized article of the uniform

E-41. **3.2.7.2.3.** Are not temporarily or permanently affixed or appended to any authorized article of the uniform.

E-42. **3.2.7.3.** The standards in subparagraph 3.2.7.2., above, are intended to serve as a basis for determining a member's entitlement under Public Law 100-80, section 508 (reference ©), to wear religious apparel with the uniform. For example, unless prohibited by subparagraph 3.2.7.6., below, a Jewish yarmulke may be worn with uniform whenever a military cap, hat or other headgear is not prescribed. A yarmulke may also be worn underneath military headgear as long as it does not interfere with the proper wearing, functioning, or appearance of the prescribed headgear.

E-43. **3.2.7.4.** Exceptions to the standards in subparagraph 3.2.7.2., above, and other special accommodations for members of particular religious groups may be granted by the Military Departments under section 4., below.

E-44. **3.2.7.5.** Whether an item of religious apparel interferes with the performance of the member's military duties depends on the characteristics of the items, the circumstances of its intended wear, and the particular nature of the member's duties. Factors in determining if an item of religious apparel interferes with military duties include, but are not limited to, whether the item may.

E-45. **3.2.7.5.1.** Impair the safe and effective operation of weapons, military equipment, or machinery.

E-46. **3.2.7.5.2.** Pose a health or safety hazard to the wearer or others.

E-47. **3.2.7.5.3.** Interfere with the wearing or proper functioning of special or protective clothing or equipment (e.g., helmets, flack jackets, flight suits, camouflaged uniforms, gas masks, wet suits, and crash and rescue equipment).

E-48. **3.2.7.5.4.** Otherwise impair the accomplishment of the military mission.

E-49. **3.2.7.6.** A complete prohibition on the wearing of any visible items of religious apparel may be appropriate under unique circumstances in which the member's duties, the military mission, or the maintenance of discipline require absolute uniformity. For example, members may be prohibited from wearing visible religious apparel while wearing historical or ceremonial uniforms; participating in review formations parades honor or color guards, and similar ceremonial details and functions.

E-50. **3.2.7.7.** The authority to approve the wearing of an item of religious apparel with the uniform, under the guidelines of this paragraph, shall be exercised at the command level specified by each Military Department. Denials of requests to wear religious apparel shall be subject to review oat the Service Headquarters level. Final review shall occur within 30 days following the date of initial denial for cases arising in the United States, and within 60 days for all other cases. Exceptions to these deadlines shall be limited to exigent circumstances, such as extended deployment. Service. Service members shall be obliged to comply with orders prohibiting the wearing of questionable items of religious apparel pending review of such order under regulations issued by the Secretaries of the Military Departments.

E-51. **3.2.7.8.** Notwithstanding paragraphs 3.2.6. and 3.2.7. above, chaplains may wear any required religious apparel or accouterments and the uniforms while conducting worship services and during the performance of rites and rituals distinct to their faith groups.

PROCEDURES 4.

E-52. **4.1.** Under rules prescribed by the Secretary of the Military Department concerned, military commanders should consider the following factors along with any other factors deemed appropriate in determining whether to grant a request for accommodation religious practices addressed in section 3., above.

E-53. **4.1.1.** The importance of military requirements in terms of individual and unit readiness, health and safety, discipline, morale, and cohesion.

E-54. **4.1.2.** The religious importance of the accommodation to the requester.

E-55. **4.1.3.** The cumulative impact of repeated accommodations of a similar nature.

E-56. **4.1.4.** Alternative means available to meet the requested accommodation.

E-57. **4.1.5.** Previous treatment of the same or similar requests, including treatment of similar requests made for other than religious reasons.

E-58. **4.2.** The factors in subsection 4.1. above are intended to promote standard procedure for resolving difficult questions involving accommodation of religious practices. In view of the different mission requirements of each command, individual consideration of specific requests for accommodation is necessary. With the exception of requests involving the wearing of visible items of religious apparel with the uniform, denials of which must be reviewed at the Service Headquarters level, the appropriate level of command for resolution of these issues shall be determined by each of the Military Departments, based on its particular requirements and circumstances.

E-59. **4.3.** When requests for accommodation are not in the best interest of the unit and continued tension between the unit's requirements and the individual's religious belief is apparent, administrative actions should be considered. These actions may include, but are not limited to, assignment, reclassification, or separation. Nothing in this Directive precludes action under the Uniform Code of Military Justice (reference (d) in appropriate circumstances.

USC, TITLE 18, POSSE COMITATUS ACT, SECTION 1385

E-60. The Posse Comitatus Act prohibits the use of the Army and Air Force to enforce civilian law (state or federal) except in cases and under circumstances expressly authorized by the constitution or act of congress. DOD policy extends this prohibition to include the Navy and Marine Corps.

E-61. This act restricts federal military law enforcement activities as follows—

- DOD military personnel do not participate directly—
  - In arrest, search and seizure, or stopping and frisking of persons, or interdiction of vessels, aircraft, or vehicles.
  - In surveillance or pursuit.
  - As informants, undercover agents, or investigators in civilian legal cases or in any other civilian law enforcement activity.
- The Posse Comitatus Act does not apply to —
  - Members of the National Guard when in Title 32 and state active duty (SAD).
  - Members of a Reserve Component when not on active duty or active duty for training.
  - DOD civilians, unless under the direct C2 of a Title 10 active duty officer.
  - Military personnel when off duty and acting solely in a private capacity.
  - Military personnel taking action for the primary purpose of furthering a military or foreign affairs function of the U.S. For example, enforcing military justice, maintaining law and order on military installations, or protecting classified materials.
  - The U.S. Coast Guard.

Appendix F

GUIDE FOR RELIGIOUS AREA/IMPACT ASSESSMENT

As the Army deploys to more areas of the world not familiar to Americans, it becomes increasingly important for chaplains to advise commanders and soldiers on the religious/spiritual aspects of the people they encounter. The term "religious/spiritual" refers to the system of beliefs and practices that give meaning and purpose to people's lives. The Unit Ministry Team (UMT) task is to understand and assess and to remain objective.

This guide gives general parameters to help the UMT fulfill its mission. Not all religious/spiritual systems will contain all of these categories. Rather they are a general matrix that will serve as a starting point to help the UMT in its assessment. The chaplain needs to first consider which of the categories are applicable to the particular society to which he has deployed. This can be done through consulting relevant literature or local subject matter experts. Having determined which categories are relevant, the chaplain can then proceed to assess how the local population understands these categories. The final step would be to covey this knowledge to commanders and soldiers as appropriate. The chaplain's answer to the questions below must address in detail how the chaplain's answer is of issue to the commander as he prepares to conduct operations.

F-1. Holy Day, Rituals and Customs—

- What Holy days and/or festivals may impact military operations?
- What are important religious "do's and don'ts" to observe?
- What are dietary restrictions?
- What are the group's distinctive symbols?
- What are the nature, frequency and traits of worship?
- What are appropriate protocols for issues related to birth/marriage/death?

F-2. Sites and Shrines—

- What are the places of worship, pilgrimage and memorial sites? Why?
- Where are the cemeteries and what is the character of their make-up?
- What are distinctive identifying characteristics (architecture, symbols, etc.)?

F-3. Primary Values—

- What are they willing to die for?

- What subjects incite an emotional response?
- What behaviors does the group reward? What are punished?
- What value is placed on women, children, ancestors, certain animals or objects?

F-4. Leadership—

- Who are the religious leaders (official and unofficial)?
- Do the religious leaders have an impact on the armed forces?
- What political influence do religious leaders have?
- What is the relationship of religious leaders to government officials?
- What do religious leaders wear to symbolize their position?
- How many leaders are there and where are they located?
- What are the leaders titles?
- How are leaders selected and trained?

F-5. Tolerance/Religious Intensity—

- What is the religious commitment in this group?
  - Nominal (in name only)
  - Mild
  - Strong
  - Radical / Fanatical
- How tolerant is the group to other parties?
- How accepting are they of conversion of their members to other groups?
- How are members disciplined?
- Can others join and quit the group easily?
- How are competing groups viewed and received?

F-6. Relationship to Society—

- How does this society relate to the religious group?
  - Stamp out the group
  - Contain the group
  - Assimilate (absorb) the group
  - Share power with the group
  - Promote pluralism with this group
- How is this group viewed?
  - Religious group
  - Secret society
  - Protest movement
  - Political party
- Does the group have a distinct subculture or communal life?
- How does the group seek to influence society?
- How do they use media resources?

F-7. Organization—

- What cell-like groups are present?
- What is the nature of hierarchy within the movement?
- What are the centers of learning?

F-8. Doctrines/Myths—

- What is/are the sources of doctrinal authority?
- What is/are the sources of ethics?
- What are their concepts of justice?
- What are their concepts of reward?
- Who are the heroes and villains? Rivalries (past, present, future)?

F-9. History/Background—

- What larger group is this group related to?
- What makes this group distinctive from the larger group?
- When did this group come into the area?

F-10. References/Sources

Appendix G

CHAPLAIN DETACHMENTS

Chaplain Detachments (CDs) provide flexibility, robustness and specific liaison functions for religious support. CDs represent a supplemental capability for conducting religious support operations and for providing/performing religious support across the full spectrum of military operations. When mobilized, these detachments increase the commander's ability to meet the religious support needs of soldiers.

There are four types of detachments:

- Chaplain Detachment–Alpha (CD-A), General Support (GS)
- Chaplain Detachment–Bravo (CD-B), (GS)
- Chaplain Detachment–Corps (CD-C), Direct Support (DS)
- Chaplain Detachment–Division (CD-D), (DS).

When mobilized, the CDs will normally come from the Reserve Components. It is important to note that these detachments are augmentation resources, not replacements for active component unit ministry teams (UMTs).

CD-A and CD-B UMTs offer maximum religious support flexibility for utilization primarily in major combat operations, but may also be utilized in homeland security and non-major combat operations. If directed by the Chief of Chaplains or FORSCOM, CD-A and CD-B UMTs may coordinate and/or provide religious support to first-responders, units without organic UMTs, DOD civilians, federal agencies, and other personnel as directed during MACA operations, WMD or other HLS incident. However, the CD-C and CD-D UMTs address specific critical religious support operational requirement shortfalls needed for the accomplishment of religious missions in the corps and division area of operations. These detachments will be aligned to support a specific corps or division. The CD-C and CD-D will have a habitual relationship for training and operational missions.

All CDs will rely on the unit to which assigned or attached for all life support and logistical support. Support may also be provided by other appropriate elements within the area of operation. This support includes, but is not limited to: legal, combat health support, finance, personnel and administrative services, field feeding, unit maintenance, and logistical support.

CHAPLAIN DETACHMENT- ALPHA (GENERAL SUPPORT)

G-1. *Mission*. The primary mission of the Chaplain Detachment – Alpha (CD-A) is to provide religious area support across the full spectrum of military operations, and spiritual reconstitution/recollection (SR2) in a Port of Debarkation (POD), Army, Theater, Corps, or equivalent separate task force. The CD-A also provides C2, operational planning and coordination support of Chaplain Detachment – Bravo (CD-B). The CD-A has the following capabilities:

- Planning, supervision, coordination, and provision of religious support for units operating in major ports of debarkation.
- Provide religious support for replacements and Force Reception, Onward movement operations in marshaling areas and ports of debarkation (PODs).
- Plan for and provide operational, specialized religious support and SR2 activities in rest, stand down, and convalescent/recovery areas. This detachment can conduct up to 2 activities per day.
- Plan for and provide area and denominational support for unit concentrations.
- Provide religious, indoctrination, and acclimatization support to incoming personnel, UMT replacements and UMTs during the reception, training, and acclimatization phase of the mobilization.
- Plan and provide religious support activities for units undergoing reconstitution.
- Provide religious support during NEOs.
- Conduct religious support operations in a JTF headquarters during operations other than war.
- Provide technical command and control (C2), supervision, control, and coordination of up to 5 CD-B UMTs.
- Plan and coordinate CD-B religious support operational missions.
- Provide the ASCC, Theater Support Command (TSC) or Corps chaplain with plans and recommendations for integrating specialized religious support and SR2 into contingency plans (COPLANs), operational plans (OPLANs) and operation orders (OPORDs) to enable support to unit/task force personnel.
- Perform other area and denominational religious support missions as required.

G-2. *Personnel—*

- Two chaplains (56A).
- Three chaplain assistants (56M).

G-3. *Assignment*. A CD-A will be assigned to an Army Service Component Command (ASCC), TSC, corps or equivalent separate task force headquarters. The unit will also be assigned as part of the port opening package of major PODs for religious support coverage of port elements. It may also be assigned to a task force or JTF headquarters.

G-4. *Employment*. A CD-A is used to augment religious support capabilities in rear areas from the division rear boundary through the communications zone in units and geographic areas where UMTs are not available or are unable to handle the religious support requirement, particularly areas such as ports of debarkation, marshaling areas, logistic support areas, combat support hospitals and other locations where there is a high concentration of soldiers. CD-A will be deployed as part of port opening packages to conduct religious support operations in major ports for port operational elements without organic UMTs. Additionally, it can be used with Force Provider or the Containerized Chapel to provide religious support to units in transit and to support personnel replacements arriving in theater. A CD-A can perform "spiritual reconstitution/recollection" activities for soldiers and members of other UMTs. If required, the CD-A can divide into two teams to provide expanded area support.

G-5. *Mobility*. The CD-A requires 100% mobility. It must have all required TOE equipment, supplies and personnel when mobilized in order to have 100% operational capability once deployed. This detachment will be required to operate independently throughout the theater, from the port(s) of debarkation (POD) to the division rear boundary while maintaining communication with the UMT of the unit to which assigned/attached. It must be able to provide technical control and coordination of up to 5 CD-Bs throughout the theaters area of operation. This detachment must be able to move all personnel and equipment in a single lift with its authorized organic vehicles.

G-6. Communications Equipment. The CD-A has a requirement to be fielded with the latest technology in both secured voice and digital data communications equipment. The CD-A will be required to communicate digitally and by voice with the unit(s) they are supporting (ASCC, TSC, or Corps or JTF HQs), CD-B UMTs, and back to home station from anywhere on the battlefield. In order for this detachment to meet its minimum mission essential wartime requirements, it must have communications equipment that is compatible and interoperable with the equipment used by the command UMT element of the unit to which assigned and units supported.

CHAPLAIN DETACHMENT-BRAVO (GENERAL SUPPORT)

G-7. *Mission*. The primary mission of the Chaplain Detachment – Bravo (CD-B) is to provide religious support by augmenting the ASCC, TSC, or Corps chaplain sections ability to provide unit and denominational religious support across the full spectrum of military operations at PODs, LSAs, base clusters, and for units without organic religious support assets or those units that deploy under concepts of split-based operations or modularity. It also provides religious support during NEOs, to units undergoing reconstitution, and mass casualties (MASCALs). The CD-B has the following capabilities:

- Provide and perform religious support activities to small units that deploy under the concepts of split-based operations or modularity and to units that do not have organic religious support assets.
- Provide and perform religious support activities for units from other services provided that, the unit to which the CD-B is assigned/attached has Army Support of other Services (ASOS) responsibility and that the unit does not have organic religious support assets.
- Provide and perform comprehensive religious support to family members prior to and during NEOs in forward deployed theatres.

- Provide and perform religious support activities for units undergoing reconstitution.
- Provide religious support to DOD civilians, contractors, and other authorized personnel, as dictated by the commander, that are in the area of operations.
- Provide SR2 when augmenting the CD-A.
- Deploy forward for immediate temporary UMT casualty replacement in divisional maneuver units as required for religious support.
- Provide other area and denominational religious support missions as required.

G-8. *Personnel—*

- One chaplain (56A).
- One chaplain assistant (56M).

G-9. *Assignment.* One CD-B will be assigned to each Corps Support Command (COSCOM) and Corps Support Group (CSG) to augment the chaplain section's ability to provide religious support within the AO. This unit can also be assigned to a TSC, ASG, or BSB.

G-10. *Employment.* Chaplain Detachment-Bravo's are used to augment religious support capabilities throughout the Theatre Army's and Corps area of operations. CD-Bs will normally be employed with a COSCOM or CSG. They can also be employed with, assigned to, or attached to Area Support Groups (ASG) or Base Support Battalions (BSB) to augment the religious support capabilities of those UMTs. Augmented UMTs will employ the CD-B to fill gaps in religious support coverage, specifically to units without organic UMTs within the assigned units area of responsibility. This would also include small units deployed under the concept of split-based operations, specifically at LSAs, and base clusters. This unit can also be utilized to assist Combat Support Hospitals (CSH) UMTs in the event of mass casualties. They can further be pushed down to the Division area of operations to support mass casualty operations when available UMT assets are unable to provide the required religious support. CD-B UMTs will not be employed as permanent UMT casualty replacements in maneuver units, but can be used as immediate temporary (usually 72 hours or less) backfill.

G-11. *Mobility.* The CD-B requires 100% mobility. It must have all required TOE equipment, supplies and personnel when mobilized in order to have 100% operational capability once deployed. This detachment will be required to operate independently throughout the theater, from the PODs to the division rear boundary. At times this detachment will also operate within the division battlespace to provide area and denominational religious support. This detachment must remain prepared to relocate on short notice (3 hours or less). It must also be able to move all personnel and equipment in a single lift with its authorized organic vehicles.

G-12. *Communications Equipment.* The CD-B has a requirement to be fielded with the latest technology in both secured voice and digital data communications equipment. The CD-B will be required to communicate digitally and by voice with the unit(s) they are supporting (ASCC, TSC, Corps, COSCOM, CSG, ASG, and BSB) and CD-A UMTs from anywhere on the battlefield. When providing temporary UMT backfill to maneuver brigades and battalions, the CD-B will be required to communicate with the division and/or maneuver brigade UMT to which they are operationally controlled to. In order for this detachment to meet its minimum mission essential wartime requirements, it must have communications equipment that is compatible with the equipment used by the command UMT element to which assigned and the unit supported.

CHAPLAIN DETACHMENT – CORPS (DIRECT SUPPORT)

G-13. *Mission.* The primary mission of the Chaplain Detachment – Corps (CD-C) is to augment the Corps Chaplain Section across the full spectrum of military operations to provide liaison with the G5/Civil Military Operation Center to facilitate the UMT supported Civil Military Operations activities within the corps on behalf of the corps chaplain. The CD-C is a religious support asset. The CD-C has the following capabilities:

- Plan for and coordinate religious support missions in support of area civil/military operations.
- Provide liaison with government, faith-based NGOs and International Organizations (IOs).
- Coordinate and monitor faith-based NGO and IO requirements and issues.
- Establish links with faith-based NGOs and IOs and with religious leaders of the host nation.
- Plan for and provide operational, specialized religious support and SR2 activities in rest, stand down, and convalescent/recovery areas.
- Provide the Corps Chaplain with plans and recommendations for contingency and reconstitution plans and support to the task force.
- Conduct specialized ministry/SR2 in the corps area at sites selected by using commanders.
- Perform other religious support missions as required.

G-14. *Personnel—*

- One chaplain (56A).
- One chaplain assistant (56M).
  - The chaplain and the chaplain assistant will have a need for specialized training in civil affairs, civil-military operations, and world religions to more effectively perform their mission. They should also have a foundational understanding and be familiar with the operations, missions, and goals of known faith-based NGOs and IOs.
  - The chaplain assistant will be required to have a 2S ASI (Battle Staff Qualified). This skill is required in order for the CD-C to work with the CMOC and the corps chaplain section in developing plans and operations. This skill is also important because it will allow the CD-C to maintain situational awareness and an understanding of the operational environment.

G-15. *Assignment.* The CD-C is assigned to all Heavy and Airborne Corps. They can further be attached or assigned to a JTF HQs or division when the division is operating as an ARFOR or JTF headquarters. The CD-C will habitually train and deploy with the corps to which assigned.

G-16. *Employment*. The CD-C will deploy with and provide habitual augmentation to the Corps chaplain section UMT. The CD-C will act as the liaison between faith-based NGO/IOs and the corps chaplain in an area of operation. The CD-C will be the principle point of contact between the Corps Chaplain and the CMOC concerning indigenous religious issues and religious support impact during Civil Military Operations missions. The CD-C may provide general religious support as the primary mission permits. The CD-C will normally co-locate with the Corps Chaplain section at the corps rear CP and operate out of either the corps chaplain cell or the CMOC. If required the CD-C can be deployed in support of a division that has ARFOR/JTF responsibility in situations where the corps HQs does not deploy. It may also be deployed forward into a division AO to conduct CA/HA and religious support liaison functions.

G-17. *Mobility*. The CD-C requires 100% mobility. It must have all required TOE equipment, supplies and personnel when mobilized in order to have 100% operational capability once deployed. This detachment will be required to operate independently throughout the corps area of operations. At times it will also be required to operate within the division area to conduct support to CMO and religious support liaison functions. This detachment will prepare to be separated from the Corps Chaplain Section for up to 72 hours. It must also be able to move all personnel and equipment in a single lift with its authorized organic vehicles.

G-18. *Communications Equipment*. The CD-C has a requirement to be fielded with the latest technology in both secured voice and digital data communications equipment. The CD-C will be required to communicate digitally and by voice with the corps and division chaplain sections from anywhere within the area of operations. They will also be required to communicate with faith-based NGO/IO representatives, host nation religious leaders, and other governmental organizations operating in the area of operations by voice communications equipment and face-to-face using either military linguists or a machine translation devices. The CD-C will also have a CONUS reach-back requirement in order to acquire information on indigenous religions and faith-based NGO/IOs from subject matter experts, databases, and the World Wide Web. In order for this detachment to meet its minimum mission essential wartime requirements, it must have communications equipment that is compatible with the equipment used by the command UMT element to which assigned and other units supported.

CHAPLAIN DETACHMENT –DIVISION (DIRECT SUPPORT)

G-19. *Mission*. The primary mission of the Chaplain Detachment – Division (CD-D) is to provide direct habitual augmentation to the Division Chaplain section. Provides assets required to enable the division chaplain section to conduct 24-hour and split-based religious support planning, coordination, and execution operations across the full spectrum of military operations. The CD-D is capable of the following:

- Provides required assets to the division chaplain section to enable sustained 24-hour and split-based operations between the DREAR and DMAIN.

- Battle track and monitor current division operations and subordinate unit and UMT locations.
- Operates battlefield-automated systems (BAS) to receive and analyze data for recommendations to the division chaplain for use in current operations and future planning of religious support.
- Collects SITREPs, reports, and requests for information (RFIs) from division CP sections/elements, corps chaplain section, brigade and separate battalion UMTs.
- Monitors locations and religious support needs and coordinates religious support for authorized civilians during operations to include: contractors, DA civilians, and interagency representatives operating in the division AO.
- Monitors location of faith-based NGO/IOs operating in the division AO and assesses religious support impact on operations.
- Perform other religious support missions as required.

G-20. *Personnel—*

- One chaplain (56A).
- One chaplain assistant (56M). The chaplain assistant will be required to have a 2S ASI (Battle Staff Qualified). This skill is required to enable the CD-D to be able to battle track and analyze current operations as a means to effectively anticipate religious support needs and develop input into current and future operational plans. It is also important because it will allow the CD-D to maintain situational awareness and an understanding of the operational environment; allowing for more accurate information exchanges to the division chaplain and staff regarding religious support operations.

G-21. *Assignment.* A CD-D will be assigned to each division.

G-22. *Employment.* A critical task for the division chaplain section is 24-hour religious support operations. The CD-D will be aligned to support and train with a specific division in order to develop a habitual relationship and operational knowledge of the division supported. Upon deployment notification of the division, the CD-D that has been assigned to the deploying division will be mobilized to deploy with that division. The CD-D will join the division chaplain section during the pre-deployment phase of the operation. This will permit the division chaplain section to conduct 24-hour and/or split-based religious support operations throughout the course of the mission and across the full spectrum of military operations. The CD-D will normally operate out of the DMAIN along with the division chaplain and NCOIC. While the CD-D can be used to perform general religious support functions, it is important to remember that the primary mission of the CD-D is to assist in performing 24-hour operations.

G-23. *Mobility.* The CD-D requires 100% mobility. It must have all required TOE equipment, supplies and personnel when mobilized in order to have 100% operational capability once deployed. This detachment will be required to have the same mobility as the division it supports. It must also be able to move all personnel and equipment in a single lift with its authorized organic vehicles.

G-24. *Communications Equipment*. The CD-D has a requirement to be fielded with the latest technology in both secured voice and digital data communications equipment. The CD-D will be required to communicate digitally and by voice with the division UMT, deputy division UMT, BDE and BN UMTs throughout the division AO. In order for this detachment to meet its minimum mission essential wartime requirements, it must have communications equipment compatible with the equipment used by the division chaplain section to which assigned and other units supported.

GLOSSARY

AA	assembly area
AAR	After Action Review
ABCA	American, British, Canadian, and Australian
AC	Active component
ACR	Armored Cavalry Regiment
ADA	air defense artillery
AGR	Active Guard Reserve
ALOC	Administrative/Logistics Operations Center
AO	area of operations
AOB	advanced operational bases
AOR	area of responsibility
AMC	Army Materiel Command
AMEDD	Army Medical Department
APOE	aerial port of embarkation
AR	Army regulation
ARFOR	Army forces
ARNG	Army National Guard
ARSOA	Army Special Operations Aviation
ARSOF	Army Special Operations Forces
ARTEP	Army Training and Evaluation Program
ASCC	Army Service Component Command
ASG	area support group
AT	annual training
AUTL	Army Universal Task List
AXP	ambulance exchange point
BAS	battalion aid station
BASOPS	base operations
BCOC	Base Cluster Operations Center
BCTP	Battle Command Training Program
BDE	brigade
BDOC	Base Defense Operations Center
BII	Basic issue items
BMO	battalion maintenance officer
BN	battalion
BOS	battlefield operating system
BSA	brigade support area
BSB	base support battalion

BUB	battle update briefings
C2	command and control
CCC (C3)	chaplain career course
C4ISR	C4 Intelligence, surveillance and Reconnaissance
CA	chaplain assistant
CA	Civil Affairs
CCH	Chief of Chaplains
CCP	Casualty Collection Point
CD	Chaplain Detachment
CD-A	Chaplain Detachment-Alpha (General Support)
CD-B	Chaplain Detachment-Bravo (General Support)
CD-C	Chaplain Detachment-Corps (Direct Support)
CD-D	Chaplain Detachment-Division (Direct Support)
CED	Critical Event Debriefing
CEOI	Communication electronic operating instructions
CFLC	Chaplain Family Life Center
CFP	Contingency Force Pool
CH	Chaplain
CHOBC	Chaplain Officer Basic Course
CISM	Critical Incident Stress Management
CJTF	combined-joint task force
CJTF	commander joint task force
CMM	Combat Medical Ministry Course
CMO	civil-military operations/officer
CMOC	Civil Military Operations Center
CMRP	Command Master Religious Plan
CMTC	Combat Maneuver Training Center
COA	course of action
CofS	Chief of Staff
COMMZ	communications zone
CONUS	Continental United States
CONUSA	Continental United States Army
COPLAN	contingency plan
COSCOM	Corps Support Command
COSR	combat operations stress response
CP	command post
CPE	Clinical Pastoral Education
CRC	Continental United States Replacement Center
CRP	Commander's Religious Program
CS	combat support
CSA	corps support area

CSB	Corps Support Battalion
CSG	Corps Support Group
CSM	Command Sergeant Major
CSS	combat service support
CSSCS	Combat Service Support Control System
CSU	Chaplain Support Unit
CT	Counter Terrorism
CTC	Combat Training Center
CTCP	Combat Trains Command Post
CZ	combat zone
DA	Department of the Army
DA	Direct Action
DACH	U.S. Army Chief of Chaplains
DASG	U.S. Army Surgeon General
DCO	Defense Coordinating Office
DCSOPS	Deputy Chief of Staff for Military Operations and Plans
DFG	Distinctive Faith Group
DFGL	Distinctive Faith Group Leaders
DISCOM	Division Support Command
DIV (IT)	Institutional Training Divisions
DIVARTY	division artillery
DLA	Defense Logistics Agency
DNVT	digital nonsecure voice telephone
DOD	Department of Defense
DOS	Days of Supply
DS	direct support
DSA	division support area
DSCP	Defense Supply Center Philadelphia
DSL	denominational service leader
DSO	domestic support operations
DTG	date time group
DTLOMS	doctrine, training, leader development, organization material, and soldier support
EAC	echelons above corps
EAD	echelons above division
EMM	Emergency Medical Ministry Course
EPW	enemy prisoner of war
FA	field artillery
FAC	family assistance center
FEMA	Federal Emergency Management Agency
FHA	foreign humanitarian assistance
FID	Foreign Internal Defense

FLOT	forward line of own troops
FORSCOM	Forces Command
FM	frequency modulation
FRAGO	fragmentary order
FRG	Family Readiness Group
FSCOORD	fire support coordinator
FSMC	forward support medical company
FTCP	Field Trains Command Post
FTX	field training exercises
GCI	Group Crisis Intervention
GS	general support
GSU	Garrison Support Unit
HA	humanitarian assistance
HHC	headquarters and headquarters company
HHT	headquarters and headquarters troop
HN	host nation
HQDA	Headquarters, Department of the Army
HSSN	home station support node
HUMINT	human intelligence
IET	initial entry training
IMA	Individual Mobilization Augmentee
IMT	Installation Ministry Team
IO	Information Operations
IPB	intelligence preparation of the battlefield
IRR	Individual Ready Reserve
ISB	intermediate staging base
JAG	Judge Advocate General
JAS	Jump Aid Station
JFC	Joint Force Commander
JFCOM	Joint Forces Command
JFLC	Joint Forces Land Component
JFLCC	Joint Forces Land Component Command
JFSOCC	Joint Force Special Operations Component Commander
JSOTF	Joint Special Operations Task Force
JOA	joint operational area
JOPES	Joint Operation Planning and Execution System
JRTC	Joint Readiness Training Center
JTF	joint task force
JSOA	Joint Special Operations Area
KIA	killed in action
LD	line of departure

LDRSHIP	loyalty, duty, respect, selfless service, honor, integrity, and personal courage
LFA	Lead Federal Agency
LL	Lessons Learned
LOGPAC	Logistics Package
LRP	logistics release point
MACA	military assistance to civil authorities
MACDIS	military assistance to civil disturbances
MACOM	Major Army Command
MASCAL	mass casualty
MDMP	Military Decision Making Process
MEDCOM	U.S. Army Medical Command
METL	mission essential task list
METT-TC	mission, enemy, terrain and weather, troops and support available, time available, and civil considerations
MIA	Missing in Action
MOB	Mobilization
MOBEX	mobilization exercises
MOOTW	military operations other than war
MOPP	mission oriented protective posture
MOS	military occupational specialty
MRI	Medical Reengineering Initiative
MRE	meal ready to eat
MS	mobilization station
MSC	Major Subordinate Command
MSCA	military support to civilian authorities
MSE	mobile subscriber equipment
MSR	main supply route
MTF	Medical Treatment Facility
MTOE	Modified Table of Organization and Equipment
NATO	North Atlantic Treaty Organization
NCO	noncommissioned officer
NCOES	Noncommissioned Officer Education System
NCOIC	noncommissioned officer in charge
NDMS	National Disaster Medical System
NEO	noncombatant evacuation operation
NG	National Guard
NGO	non-governmental organization
NMS	National Military Strategy
NOK	next of kin
NOVA	National Organization for Victim Assistance
NSN	national stock number

OCCH	Office of the Chief of Chaplains
OCOKA	Observation and fields of fire, cover and concealment, obstacles, Key terrain, and avenues of approach
OCONUS	Outside the Continental United States
ODA	Operational Detachment Alpha
ODSS	Offensive, Defensive, Stability and Support
OES	Officer Education System
OOTW	operations other than war
OPCON	operational control
OPLAN	operations plan
OPORD	operations order
OPSEC	operations security
OPTEMPO	operation tempo
OVM	operator's vehicle material
PAO	Public Affairs Office/Officer
PBO	Property Book Officer
PCI	precombat inspection
PCS	permanent change of station
PERSCOM	Personnel Command
PHA	Personnel Holding Area
PIES	proximity, immediacy, expectancy, simplicity
PMCS	preventive maintenance checks and services
POD	port of debarkation
POE	port of embarkation
POM	preparation of overseas movement
POTF	PSYOP Task Force
POTG	PSYOP Task Group
POW	prisoner of war
PROFIS	Professional Officer Filler System
PSYOP	psychological operations
RAA	Religious Area Analysis
RAOC	Rear Area Operation Center
RC	Reserve component
RCP	Rear Command Post
RIP	Ranger Indoctrination Program
RFI	request for information
RMST	Religious Ministry Support Team
ROE	rules of engagement
ROP	Ranger Orientation Program
RPP	Religious Preference Profile
RS	religious support
RSC	Reserve Support Command

RSE	religious support estimate
RSOI	reception, staging, onward movement, and integration
RSP	religious support plan
RTF	Reconstitution Task Force
RTOC	Rear Tactical Operation Center
S1/G1	Adjutant/Personnel Officer
S2/G2	Intelligence Officer
S3/G3	Operations and Training Officer
S4/G4	Logistics Officer
S5/G5	Civil Military Operations officer
S6/G6	Signal operations officer
SACRED	spirituality, accountability, compassion, religious leadership, excellence, and diversity
SAD	state active duty
SEAD	Suppression of Enemy Air Defenses
SF	Special Forces
SFOB	Special Forces Operational Base
SFSC	Special Forces Staff Course
SIDPERS	Standard Installation/Division Personnel System
SIGO	signal officer
SITREP	situation report
SJA	Staff Judge Advocate
SMART-B	Special Medical Augmentation Response Team-Burn
SMART-CB	Special Medical Augmentation Response Team-Chemical/Biological
SMART-HS	Special Medical Augmentation Response Team-Health Systems Assessments and Assistance
SMART-MC3T	Special Medical Augmentation Response Team- Medical Command, Control, Communications, Telemedicine
SMART-PC	Special Medical Augmentation Response Team-Pastoral Care
SMART-PM	Special Medical Augmentation Response Team-Preventive Medicine/Disease Surveillance
SMART-SM	Special Medical Augmentation Response Team-Stress Management
SMART-TCC	Special Medical Augmentation Response Team-Trauma/Critical Care
SMART-V	Special Medical Augmentation Response Team-Veterinary
SO	special operations
SOCOM	Special Operations Command
SOF	Special Operation Forces
SOFA	status of forces agreement
SOFCOM	United States Special Operations and Forces Command
SOI	special operating instruction
SOP	standard operating procedures
SPOE	seaport of embarkation

SR	Special Reconnaissance
SR	supply route
SR2	spiritual reconstitution/recollection
SRP	Soldier Readiness Program
SSC	small-scale contingency
SSCO	small-scale contingency operations
STARC	State Area Command, ARNG
TAA	tactical assembly area
TACON	tactical control
TDA	Table of Distribution and Allowances
TF	task force
TF A/L	task force (Administrative/Logistics)
TOA	transfer of authority
TOC	tactical operations center
TSC	Theater Support Command
TSOP	tactical standard operating procedure
TTP	tactics, techniques and procedures
UCMJ	uniformed code of military justice
UMCP	Unit Maintenance Collection Point
UMT	Unit Ministry Team
USACAPOC (A)	United States Army Civil Affairs and Psychological Operations Command (Airborne)
USAR	U.S. Army Reserve
USARC	United States Army Reserve Command
USASOC	United States Army Special Operations Command
USC	United States Code
USDB	United States Disciplinary Barracks
USSOCOM	United States Special Operations Command
UW	unconventional warfare
WIA	wounded in action
WMD	weapons of mass destruction
XO	executive officer

Bibliography

DOCUMENTS NEEDED

These documents should be available to the intended users of this publication.

Except where otherwise indicated, the following DA forms are available on the Army Electronic Library (AEL) CD-ROM (EM 0001) and the USAPA web site (www.usapa.army.mil); DD forms are available from the OSD web site http://www.dior.whs.mil/icdhome/ddeforms.htm

DA Form 7462, Religious Support Matrix, September 2002.

DD Form 1610, Request and Authorization for TDY Travel of DOD Personnel, January 2001.

JOINT AND MULTISERVICE PUBLICATIONS

Most joint publications are online at http://www.dtic.mil/doctrine/

DOD Directive 1300.6, *Conscientious Objectors*, 20 August 1971.

DOD Directive 1300.17, *Accommodation of Religious Practices Within the Military Services*, 3 February 1988.

DOD Directive 1332.14, *Enlisted Administrative Separations*, 21 December 1993.

JP 1-0, *Doctrine for Personnel Support to Joint Operations*, 19 November 1998.

JP 1-05, *Religious Ministry Support for Joint Operations*, 26 August 1996.

JP 3-0, *Doctrine for Joint Operations*, 10 September 2001.

JP 3-05, *Doctrine for Joint Special Operations*, 17 April 1998.

JP 3-07.3, *Joint Tactics, Techniques, and Procedures for Joint Special Operations Task Force Operations*, 19 December 2001.

JP 3-08, *Interagency Coordination During Joint Operations*, Volumes I and II, 9 October 1996.

JP 3-10, *Joint Doctrine for Rear Area Operations*, 28 May 1996.

JP 5-00.2, *Joint Task Force (JTF) Planning Guidance and Procedures*, 13 January 1999.

ARMY PUBLICATIONS

Most Army doctrinal publications are online at http:www.adtdl.army.mil/atdls.html or at http://www.lia.army.mil/dpmsh/pubs_rev_dt.html

AR 165-1, *Chaplain Activities in the United States Army*, 27 February 1998.

AR 380-5, *Department of the Army Information Security Program*, 29 September 2000.

AR 600-20, *Army Command Policy*, 13 May 2002.

AR 680-29, *Military Personnel – Organization and Type of Transactions Codes*, 1 March 1989.

DA PAM 165-3, *Chaplain Training Strategy*, 1 September 1998.

FM 1 (100-1), *The Army*, 14 June 2001.

FM 1-111, *Aviation Brigades*, 27 October 1997.

FM 3-0 (100-5), *Operations*, 14 June 2001.

FM 3-90 (100-40), *Tactics*, 4 July 2001.

FM 3-90.3 (71-3), *The Mounted Brigade Combat Team*, 1 November 2001.

FM 7-0 (25-100), *Train the Force*, 22 October 2002.

FM 7-20, *The Infantry Battalion*, 6 April 1992.

FM 7-30, *The Infantry Brigade*, 3 October 1995.

FM 8-51, *Combat Stress Control in a Theater of Operations Tactics, Techniques and Procedures*, 29 September 1994.

FM 16-1, *Religious Support*, 26 May 1995.

FM 22-51, *Leaders' Manual for Combat Stress Control*, 29 September 1994.

FM 22-100, *Army Leadership*, 31 August 1999.

FM 25-101, *Battle Focused Training*, 30 September 1990.

FM 55-30, *Army Motor Transport Units and Operations*, 27 June 1997.

FM 71-2, *The Tank and Mechanized Infantry Battalion Task Force*, 27 September 1988.

FM 71-100, *Division Operations*, 28 August 1996.

FM 100-6, *Information Operations*, 27 August 1996.

FM 100-7, *Decisive Force: The Army in Theater Operations*, 31 May 1995.

FM 100-8, *The Army in Multinational Operations*, 24 November 1997.

FM 100-9, *Reconstitution*, 13 January 1992.

FM 100-10-2, *Contracting Support on the Battlefield*, 4 August 1999.

FM 100-11, *Force Integration*, 15 January 1998.

FM 100-15, *Corps Operations*, 29 October 1996.

FM 100-17-3, *Reception, Staging, Onward Movement, and Integration*, 17 March 1999.

FM 100-19, *Domestic Support Operations*, 1 July 1993.

FM 100-20, *Military Operations in Low Intensity Conflict*, 5 December 1990.

FM 100-21, *Contractors on the Battlefield*, 26 March 2000.

FM 100-23, *Peace Operations*, 30 December 1994.

FM 101-5, *Staff Organization and Operations*, 31 May 1997.

FM 101-5-2, U.S. Army Report and Message Formats, 29 June 1999

OTHER PUBLICATIONS

Army Vision 2010

Chaplain Assistant Concept Paper, 2000

Chaplain Support Force XXI White Paper

Chief of Chaplains MSCA Concept Paper, September 2000

The Chief of Chaplain's UMT Crisis Assets Handbook, October 2000

Concept for Future Joint Operations, May 1997

Joint Vision 2010

Joint Vision 2020 –http://www.dtic.mil/jv2020

Military Rules of Evidence 503 UCMJ

National Military Strategy

A National Security Strategy for a New Century, May 1997

The Quadrennial Defense Review Report, September 2001.

Public Law 98-525

Public Law 100-180

Textbook of Military Medicine: War Psychiatry, Part I, Warfare, Weaponry and the Casualty Office of the Surgeon General of the United States Army, Falls Church, VA, 1995.

United States Code, Title 10, Armed Forces http://uscode.house.gov/title_10.htm

United States Code, Title 18, Crimes and Criminal Procedures: http://uscode.house.gov/title_18.htm

U.S. Army Chaplaincy Strategic Plan FY 2000-FY 2005

PREFACE

Combat Stress is the mental, emotional or physical tension, strain, or distress resulting from exposure to combat and combat-related conditions. Controlling combat stress is a command responsibility. In terms of Service members lost from action and reduced performance, combat stress seriously affects mission accomplishment. It is a leader's responsibility to take action to strengthen Service members' tolerance to combat stress and manage it in his or her unit.

Combat stress reactions are the result of exposure to the same conditions during military actions that cause physical injury and disease in battle or its immediate aftermath, and many combat stress reactions occur in persons who are also wounded or ill with disease. Rates of combat stress casualties vary greatly, with higher ratios during lengthy periods of intense combat. In Okinawa 1945, during a peak month of battle, the combat stress casualties among Marine Forces were reported as high as one for every two wounded in action (WIA). Under less lengthy periods, as suggested by data acquired from the Israeli Defense Forces fighting in Lebanon 1982, the ratio of combat stress casualties to WIA in small units can be as high as one to one. In the past, we have generally suffered as many as one battle stress casualty for every three to five WIA in heavy fighting. However, highly trained units with strong leadership and high esprit de corps have fewer combat stress casualties.

While this manual focuses on combat-induced stress reactions, it is important to emphasize that "combat stress" is not restricted only to combat, but may also arise from combat-like conditions present during military operations other than war. In an area of operations characterized by continuous action and high danger, our forces may experience high rates of stress casualties unless small-unit leaders are trained and prepared to manage stress.

This publication is written to inform small-unit leaders of stress characteristics and management techniques in order to *prevent, reduce, identify,* and *treat* combat stress reactions in the Service member's own unit to the maximum extent possible. A significant part of training is learning to control and cope with stress. Leaders must learn to cope with their own stress and then assist junior personnel in managing their stress. The application of combat stress management techniques helps conserve fighting strength and provides one more step toward achieving success.

Table of Contents

Chapter 2. Combat Stress Prevention, Management, and Control

Chapter 3. Command Leadership Actions

Chapter 4. Sleep Deprivation

Appendices

Chapter 1

Combat Stress Identification

1001. INTRODUCTION

The Marine Corps' success as a fighting force is dependent on leadership that maintains a balanced focus between mission accomplishment and troop welfare. The small-unit leader is the key to building and maintaining high unit morale and peak efficiency. He achieves this in part by knowing his troops and understanding their strengths and weaknesses. To maintain that same level of morale and efficiency in combat, the small-unit leader must understand how to recognize, prevent, and even personally contend with reactions to combat stress when it occurs in his unit. If a condition accounted for as many casualties in combat and the condition was at least partially preventable, the prudent combat leader would be interested in knowing more about it. Combat stress reaction(s), also called battle fatigue, is that condition. It has the potential to disable the most courageous Service member and influence the success or failure of a unit in accomplishing its mission.

1002. HISTORY

During the 1942-45 period in the European Theater, there was a ratio of one combat stress casualty for every three WIA. In a month of especially horrible, continuous fighting in Okinawa in 1945, the 6th Marine division had one stress casualty for every 1.8 WIA. However, the airborne divisions in Europe never had more than one for ten WIA, and usually less, even though they experienced very high casualties in some battles.

To accomplish a mission successfully, planners must use some guidelines to estimate losses from combat stress. What are the reasons for the tremendous range between high and low battle stress casualty rates? While the answer to this question is complex, it is clear that better-trained troops have fewer killed and wounded, and proportionally fewer stress casualties. Service members—especially leaders—can learn to recognize the symptoms and prevent or reduce the disruptive effects of combat stress.

1003. REACTIONS TO COMBAT STRESS

Service members exposed to danger experience physical and emotional reactions that are not present under more tranquil circumstances. Some reactions sharpen abilities to survive and win; other reactions may produce disruptive behaviors and threaten individual and unit safety. These adverse behaviors are collectively called *combat stress reaction*. The operative word is "behaviors." People in combat experience a range of emotions, but their behavior influences immediate safety and mission success.

Combat and combat-related military missions can also impose combinations of heavy physical work; sleep loss; dehydration; poor nutrition; severe noise, vibration, and blast; exposure to heat, cold or wetness; poor hygiene facilities; and perhaps exposure to infectious diseases, toxic fumes or substances. These, in combination with other influences—such as concerns about problems back home—affect the ability to cope with the perception of danger, and diminish the skills needed to accomplish the mission. Environmental stressors often play an important part in causing the adverse or disruptive combat stress reaction behaviors. The leader must work to keep each Service member's perception of danger balanced by the sense that the unit has the means to prevail over it. The leader must keep himself and his unit working at the level of stress that sustains performance and confidence. When troops begin to lose confidence in themselves and their leader, adverse stress reactions are most likely to occur.

It is important for the small-unit leader to recognize these adverse behaviors at the onset in order to intervene promptly for the safety and benefit of individual Service members and the unit. These behaviors may take many forms and can range from subtle to dramatic. Any Service member who shows persistent, progressive behavior that deviates from his baseline behavior may be demonstrating the early warning signs and symptoms of a combat stress reaction. Trying to memorize every possible sign and symptom is less useful to prompt diagnosis than to keep one simple rule in mind: *Know your troops, and be alert for any sudden, persistent or progressive change in their behavior that threatens the functioning and safety of your unit.*

1004. OBSERVING AND RECOGNIZING COMMON REACTIONS TO COMBAT STRESS

Ranges of fatigue, fear, anxiety, and depression affect most Service members in combat and in some military operations other than war. Mild stress reaction may be signaled by changes in behavior and only be discernible by the person himself or by close comrades. The unit leader and medical personnel depend on information from the Service member or his comrades for early recognition of combat stress reactions to provide prompt and appropriate help. Table 1-1 lists some mild stress reactions.

Table 1-1. Mild Stress Reactions.

PHYSICAL	EMOTIONAL
Trembling	Anxiety, indecisiveness
Jumpiness	Irritability, complaining
Cold sweats, dry mouth	Forgetfulness, inability to concentrate
Insomnia	Nightmares
Pounding heart	Easily startled by noise, movement, and light
Dizziness	Tears, crying
Nausea, vomiting or diarrhea	Anger, loss of confidence in self and unit
Fatigue	
"Thousand-yard" stare	
Difficulty thinking, speaking, and communicating	

Severe stress reactions may prevent the individual from performing his duties or create a concern for personal safety or the safety of others. A variety of more serious reactions or warning signs are listed in Table 1-2. These do not necessarily mean that the person must be relieved from duty, but warrant immediate evaluation and help by the small-unit leader.

Table 1-2. Severe Stress Reactions.

PHYSICAL	EMOTIONAL
Constantly moves around	Talks rapidly and/or inappropriately
Flinches or ducks at sudden sound and movement	Argumentative; acts recklessly
Shakes, trembles	Indifferent to danger
Cannot use part of body (hand, arm leg) for no apparent physical reason	Memory loss
Inability to see, hear or feel	Stutters severely, mumbles or cannot speak at all
Is physically exhausted; cries	Insomnia; severe nightmares
Freezes under fire or it totally immobile	Sees or hears things that do not exist
States vacantly, staggers or sways when standing	Has rapid emotional shifts
Panics, runs under fire	Socially withdrawn
	Apathetic
	Hysterical outbursts
	Frantic or strange behavior

Fatigue

The most common stress reactions include:

- Slow reaction time.
- Difficulty sorting out priorities.
- Difficulty starting routine tasks.
- Excessive concern with seemingly minor issues.
- Indecision, difficulty-focusing attention.
- A tendency to do familiar tasks and preoccupation with familiar details. These reactions may reach a point where the person becomes very passive, or wanders aimlessly.
- Loss of initiative with fatigue and exhaustion.

Muscular Tension

Headaches, Backache

Muscular tension increases strain on the scalp and spine and often leads to headaches, pain, and cramps.

Inability to Relax

Prolonged muscular tension wastes energy and leads to fatigue and exhaustion. Muscles must relax periodically to enable free blood flow, waste product flushing, and nutrient replenishment.

Shaking and Tremors

Mild Shaking

During incoming rounds, the individual may experience mild shaking. This symptom appears and disappears rapidly and is considered a normal reaction to conditions of great danger.

Marked or Violent Shaking

A common post-battle reaction, marked or violent shaking can be incapacitating if it occurs during the action. If shaking persists long after the precipitating stimulus ceases, or if there was no stimulus, the individual should be checked by medical personnel.

Perspiration

It is normal to experience either mild or heavy sweating or sensations of chilliness under combat stress.

Digestive and Urinary Systems

Nausea and Vomiting

"Butterflies in the stomach" is a common stress feeling. Vomiting may occur as a result of an extreme experience like that of a fire-fight, shelling, etc., or in anticipation of danger.

Loss of Appetite

Appetite loss may result as a reaction to stress. It becomes a significant problem if rapid weight loss occurs, or the person does not eat a sufficiently balanced diet to keep his muscles and brain supplied for sustained operations.

Abdominal Distress

Acute abdominal pain ("knotted stomach," "heartburn") may occur during combat. Persistent and severe abdominal pain is a disruptive reaction, and may indicate a medical condition.

Frequent Urination

Frequent urination may occur, especially at night.

Incontinence

During extremely dangerous moments, inability to control bowel and/or bladder functions may occur. Incontinence is embarrassing, but it is not abnormal under these circumstances.

Circulatory and Respiratory Systems

Heart Palpitations

Rapid heartbeat, a sense of pressure in the chest, occasional skipped beats, and sometimes chest pains are common with anxiety or fear. Very irregular heartbeats need to be checked by medical personnel.

Hyperventilation

Hyperventilation is identified by rapid respiration, shortness of breath, dizziness, and a sense of choking. It is often accompanied with tingling and cramping of fingers and toes. Simple solutions are increased exercise and breathing with a paper bag over the nose and mouth; or breathing slowly using abdominal muscles (called "abdominal breathing").

Faintness and Giddiness

These reactions occur in tandem with generalized muscular weakness, lack of energy, physical fatigue, and extreme stress. Brief rest should be arranged, if possible.

Sleep Disturbance

Difficulty Falling Asleep

Sometimes a Service member who has experienced intense battle conditions often cannot fall asleep even when the situation permits, or, when he does fall asleep, he frequently wakes up and has difficulty getting back to sleep.

Nightmares

Terror dreams, battle dreams, and nightmares of other kinds cause difficulty in staying asleep. Sleep disturbances in the form of dreams are part of the coping process. This process of *working through* combat experiences is a means of increasing the level of tolerance of combat stress. The individual may have battle-related nightmares or dreams that a close relative (spouse, parent) or another person important in his life has been killed in the battle. As time passes, the nightmares usually occur with less intensity and less frequency. In some cases, a Service member, even when awake, may re-experience the memory of the stressful experience as if it were recurring (called a "flashback"). This is usually triggered by a smell, sound or sight, and is not harmful as long as the Service member realizes it is only a memory and does not react inappropriately or feel overwhelmed. However, if it happens frequently or is very distressing, help should be sought from the chaplain or medical.

Restless Sleep

When a person is asleep, the sleep is not restful sleep if the person is constantly being half-wakened by noise, movement, or other stimuli. Heavy snoring often indicates poor quality sleep. The individual wakes up as tired as when he went to sleep. Finding a more comfortable position, away from distractions, can help.

Excessive Sleep

Individuals exhibiting a need for excessive sleep may be exhibiting symptoms of combat stress; however, excessive sleep is also a sign of substance abuse or depression (Persistent insomnia is a more common indicator of possible depression.)

Visual/Hearing Problems and Partial Paralysis

Stress-related blindness, deafness, loss of other sensations, and partial paralysis are not true physical injuries, but physical symptoms that unconsciously enable the individual to escape or avoid a seemingly intolerably stressful situation. These symptoms can quickly improve with reassurance and encouragement from comrades, unit medical personnel or battalion physician. If they persist, the physician must examine the individual to be sure there is not a physical cause; for example, laser range finders can cause temporary or partial blindness, and nearby explosions can cause ear damage. Individuals with these physical conditions are unaware of the causative relationship with their inability to cope with stress. They are honestly concerned with their physical symptoms and want to get better. They are willing to discuss them, and do not mind being examined. This is contrary to "malingerers" faking a physical illness, who are often reluctant to talk, or over-dramatize their disability and refuse an examination.

Visual problems include blurred vision, double vision, difficulty in focusing, or total blindness. Hearing problems include inability to hear orders and/or nearby conversations or complete deafness occurs. Paralysis or loss of sensation is usually confined to one arm or leg. Prickling sensations or rigidity of the larger joints occur. However, temporary complete immobility (with normal breathing and reflexes) can occur. If these reactions do not recover quickly with immediate reassurance, care must be taken in moving the casualty to medical for an evaluation to avoid making a possible nerve or spinal cord injury worse.

Bodily Arousal

Not all emotional reactions to stress are necessarily negative. For example, the body may become aroused to a higher degree of awareness and sensitivity.

Threat

In response to threat, the brain sends out chemicals arousing the various body systems. The body is ready to fight or take flight. The alerting systems of the experienced combat veteran become finely tuned, so that he may ignore loud stimuli that pose no danger (such as the firing of nearby friendly artillery). However, he may awaken from sleep at the sound of an enemy mortar being fired and take cover before the round hits. The senses of vision and smell can also become very sensitive to warning stimuli. The Service member may instantly focus and be ready to react.

Hyperalert

Hyperalert refers to being distracted by any external stimuli that might signal danger, and overreacting to things that are, in fact, safe. The hyperalert Service member is not truly in tune with his environment, but is "on a hair trigger." He is likely to over-react, misinterpret reassuring information as threats, and react without adequate critical thinking. Consequences can range from firing at an innocent noise to designating as hostile an innocent target.

Startle Reactions

Startle reactions are part of an increased sensitivity to minor external stimuli (on-guard reactions). Leaping, jumping, cringing, jerking, or other forms of involuntary self-protective motor responses to sudden noises are noted. The noises are not necessarily very loud. Sudden noise, movement, and light cause startle reactions; for example, unexpected movement of an animal (or person) precipitates weapon firing.

Anxiety (Fear of Death, Pain, and Injury)

Fear of death, pain, and injury causes anxiety reactions. After witnessing the loss of a comrade in combat, a Service member may lose self-confidence and feel overly vulnerable or incapable. The death of a buddy leads to serious loss of emotional support. Feelings of "survivor guilt" are common. The survivors each brood silently, second-guessing what they think they might have done differently to prevent the loss. While the Service member feels glad he survived, he also feels guilty about having such feelings. Understanding support and open grieving shared within the unit can help alleviate this.

Irritability

Reaction

Mild irritable reactions range from angry looks to a few sharp words, but can progress to more serious acts of violence. Mild irritability is exhibited by sharp, verbal overreaction to normal, everyday comments or incidents; flare-ups involving profanity; and crying in response to relatively slight frustrations.

Explosion of Aggressive Behavior

Sporadic and unpredictable explosions of aggressive behavior (violence) can occur with little or no provocation. For example, a

Service member tries to pick a fight with another Service member. The provocation may be a noise, such as the closing of a window, an accidental bumping or normal, verbal interaction.

Short Attention Span

Persons under stress have short attention spans. They find it difficult to concentrate. Short attention span causes a Service member to have difficulty following orders. The Service member does not easily understand what others are saying. The person has difficulty following directions, aiding others or performing unfamiliar tasks.

Depression

People respond to stress with protective defensive reactions against painful perceptions. Emotional dulling or numbing of normal responsiveness is a result. The reactions are easily observed changes from the individual's usual self.

Low Energy Level

The observer notes decreased effectiveness on the job, decreased ability to think clearly, excessive sleeping or difficulty falling asleep, and chronic tiredness. Emotions such as pride, shame, hope, grief, and gratitude no longer matter to the person.

Social Withdrawal

A Service member is less talkative than usual, shows limited response to jokes or cries. He is unable to enjoy relaxation and companionship, even when the tactical situation permits.

Change in Outward Appearance

If the Service member is in a depressed mood, he may be observed to exhibit very little body movement and an almost expressionless (mask-like) face.

Substance Abuse

Some Service members may attempt to use substances such as alcohol or other drugs as a means of escaping combat stress. The use of substances in a combat area makes some Service members less capable of functioning on the job. They are less able to adapt to the tremendous demands placed on them in combat.

Loss of Adaptability

Less common reactions include uncontrolled emotional outbursts such as crying, yelling, or laughing. Some Service members may become withdrawn, silent, and try to isolate themselves. Uncontrolled reactions can appear singly or in combination with a number of other symptoms. In this state, the individual may become restless, unable to keep still, and move aimlessly about. He may feel rage or fear, which he demonstrates by aggressive acts, angry outbursts or irritability.

Disruptive Reactions

Service members with disruptive, combat stress reactions—

- **Cannot function on the job.** In some cases, stress produces symptoms often associated with head injuries. For example, the person may appear dazed and may wander around aimlessly. He may appear confused and disoriented, and exhibit either a complete or partial memory loss. Service members exhibiting this behavior should be removed from duties until the cause for this behavior can be determined.

- **Compromise their own safety.** In a desperate attempt to escape the danger that has overwhelmed him, an individual may panic and become confused. The term *panic run* refers to a person rushing about without self-control. In combat, such a Service member easily compromises his safety and could possibly get killed. His mental ability becomes impaired to the degree that he cannot think clearly or follow simple commands. He stands up in a firefight because his judgment is clouded and he cannot understand the likely consequences of his behavior. He loses his ability to move and seems paralyzed.
- **Compromise the safety of others.** If panic is not quelled early, it can easily spread to others. A person in panic is virtually out of control and needs to be protected from himself. More than one person may be needed to exert control over the individual experiencing panic. However, it is also important to avoid threatening actions, such as striking him.

Referral to the Chaplain, Medical Officer or Combat Stress Control Team

Although the more serious or warning behaviors described in the preceding paragraphs usually diminish with help from comrades and small unit leaders, and time, some do not. An individual usually improves when able to get warm food, rest and an opportunity to share his feelings with comrades or small unit leader. If the symptoms endanger the individual, others or the mission, or if they do not improve within a day or two, or seem to worsen, get the individual to talk with the unit chaplain or medical officer. Access to mental health/combat stress control specialists may be sought, if available. **Do not** wait too long to see if the experience is better with time. Specialized training is not required to recognize severe stress reactions. The small-unit leader can usually determine if the individual is not performing his duties normally, not taking care of himself, behaving in an unusual fashion, or acting out of character.

1005. SUICIDE AWARENESS

Some behaviors and symptoms previously described are not only signs of stress reaction, but can also signal potential suicide risks. Service members must be ever vigilant for the signs and signals of a potential threat of suicide given by their fellow Service members. Appendix A provides useful tools for identifying these signals.

Chapter 2

Combat Stress Prevention, Management, and Control

2001. PREVENTING AND MANAGING COMBAT STRESS

The same leadership skills that apply to troop welfare and warfighting can effectively reduce or prevent combat stress reactions. Leaders should take preventive actions and address stress symptoms as they appear. Ignoring the early warning signs can increase the severity of stress reactions. Positive action to reduce combat stress also helps Service members cope with normal, everyday situations and makes them less likely to experience harmful combat stress reactions. Table 2-1 lists stress management techniques.

2002. STRESS-REDUCTION TECHNIQUES FOR LEADERS

To reduce stress, the leader should—

- Lead by inspiration, not fear or intimidation.
- Initiate and support stress management programs.
- Provide information to focus stress positively.
- Ensure each Service member has mastered at least two stress coping (relaxation) techniques, a slow one for deep relaxation and a quick one for on the job.

Table 2-1. Stress Management Techniques.

Assure every effort is made to provide for the troops' welfare.
Instill confidence in each Service member and his equipment, unit, and leadership.
Be decisive and assertive; demonstrate competence and fair leadership.
Provide sleep and/or rest, especially during continuous operations, whenever possible.
Ensure sleep for decisionmaking personnel.
Set realistic goals for progressive development of the individual and team.
Systematically test the achievement of these goals.
Recognize that battle duration and intensity increase stress.
Be aware of environmental stressors such as light level, temperature, and precipitation.
Recognize that individuals and units react differently to the same stress.
Learn the signs of stress in yourself and others.
Recognize that fear is a normal part of combat stress.
Rest minor stress casualties briefly, keeping them with their unit.
Be aware of background stress sources prior to combat; e.g., family concerns and/or separation, economic problems.
Provide an upward, downward, and lateral information flow to minimize stress due to a lack of communication.

Table 2-1. Stress Management Techniques (Continued).

Practice stress control through cross-training, task allocation, tasks matching, and task sharing.
Look for stress signs and a decreased ability to tolerate stress.
Practice and master stress-coping techniques.
Face combat stress; it is unhealthy to deny the stresses of combat.

- Look out for Service members' welfare.
- Communicate with Service members personally to learn of stressors and detect signs of stress.
- Understand that stress in response to threatening or uncertain situations is normal.
- Create a spirit to win under stress.
- Act as role model for self-control of stress reactions.

Realistic training is the primary stress-reduction technique. It assures Service members' maximum confidence in their skills and belief that their leaders are doing their best for them. Since the basic necessities of life assume even greater importance on the battlefield, leaders should:

- Ensure personnel are properly trained.
- Ensure training includes understanding of combat stress and how to deal with it.
- Place welfare of subordinates before personal welfare, but keep them capable.
- Ensure Service members get as much rest as possible.

- Ensure the best possible shelters are available.
- Keep Service members well supplied with food, water, and other essentials.
- Provide mail, news, and information avenues.
- Provide the best medical, logistical, and other support.
- Maintain high morale, unit identity, and esprit de corps.
- Keep the same unit members together.
- Assure that experienced unit members take care of and teach new members.

The unit's encouragement and support of Service members' efforts to cope with stress have a decisive effect. Unit actions can determine if Service members will endure combat exposure and accomplish their mission. Stress-coping indoctrination is part of every unit's combat training and is followed by a program of action. Controlling stress requires practice. Programs are tailored to individual units and improve with experience. Coping with stress is practiced under conditions as similar to combat as possible. A unit's ability to cope with stress in combat depends primarily on how rigorously and realistically training has been conducted.

2003. PERFORMANCE DEGRADATION PREVENTIVE MEASURES

Every Service member, team, and unit must learn to effectively sustain performance in continuous operations. This requirement applies especially to leaders. While it is an important ingredient, the determination to endure does not ensure effectiveness. Gaining the required capability goes beyond a high level of proficiency in combat skills and technical specialties. It means learning to identify the adverse conditions of continuous operations, cope with them, and overcome their effects. It also means learning how to slow the rate of performance degradation. Units, leaders, and personnel must prepare and execute plans and train to sustain performance. Adverse conditions progressively degrade Service member effectiveness. Fortunately, long-term remedies exist for slowing the rate of performance decline. These remedies, which must be introduced prior to combat, include safety, food intake, combat load, and physical fitness.

Safety

Safety, which encompasses such factors as using proper lifting techniques and staying alert and careful, is influenced by fatigue. Overtired Service members are more vulnerable to injury than those who are rested. After 72 hours of continuous combat, the tendency to seek shortcuts is very strong, and accident rates increase 50 percent. Fatigue affects all military systems, but it is especially hazardous when weapon systems are involved. Catastrophic accidents can occur when fatigued (and under-experienced) crews man weapon systems. Ways to safeguard Service members include developing and following safety standing operating procedures and increasing supervision during extended operations.

Food Intake

If Service members are too busy, stressed or tired to eat adequate rations during continuous operations, their caloric intake will be reduced. This may lead to both physical and mental fatigue and degraded performance. For example, in accidents judged to involve aviator fatigue, there is some indication that before the accidents occurred, the pilots had irregular eating schedules or missed one or more meals. In field tests done by the U.S. Army Natick Research, Development, and Evaluation Center, Natick,

MA, some personnel subsisting solely on meals, ready to eat (MREs) lost weight over just a few weeks. Leaders must encourage troops to eat all of the main items in their MREs, not just the candy, to get balanced nutrition. In various other field tests conducted during continuous operations, it was discovered that meals were frequently delivered late or missed altogether. Although the relationship between performance and nutrition is not clear, eating regularly is important in continuous operations, as well as providing hot meals at assigned times or when the workload has been sustained boosts morale.

Good nutrition is important. An inadequate diet degrades performance, reduces resistance to disease, and prolongs recuperation from illness and injury. When food and water become available in continuous operations, leaders must provide and maintain a supply of food that has the nutritional value commensurate with the physical activity and stress of battle. They must remind and encourage Service members to eat and drink properly. The excitement, stress, and rapid pace of events associated with field preparations can cause Service members to forget to drink liquids. Thus, they enter the early part of the field scenario inadequately hydrated. Dehydration may result, especially if the early scenario calls for assault of a position or rapid air/land deployment. Contributing to developing dehydration is the relative lack of moisture in MREs and other packets. In addition, Service members experiencing dehydration lose their appetite and reduce their food intake. This, in combination with dehydration, leads to degraded performance. Leaders must reemphasize drinking regimens to ensure that Service members are properly hydrated going into battle. Leaders must remind Service members to drink liquids in both hot and cold climates and must monitor fluid intake. If personnel drink only when thirsty, they will become dehydrated.

Combat Load

In combat, the load carried by a Service member significantly exceeds optimum recommended weights. In the case of a light infantry Service member, the combat load is as much as double the recommended load. Physical conditioning cannot compensate for this degree of excess. Service members tire faster and, in continuous combat, recovery from fatigue becomes more time-consuming.

When the Service member must carry excessive amounts of equipment, the effects of stress and lack of rest are magnified. Employing a load echelonment concept must be considered to ease the strain on Service members. In this concept, the unit separates an individual's equipment into two loads—fighting and existence. As the unit closes the objective, the heavier existence load is dropped and the Service member continues with the lighter fighting load.

Physical Fitness

Good physical conditioning delays fatigue, builds confidence, and shortens recovery times from illness and injury. It also prepares individuals to better cope with the physiological demands of stress. Service members in top physical condition can better control their internal physiological functions, which will improve their overall performance. Physical fitness—including aerobic fitness, muscular strength, and endurance—must be developed in all Service members to strengthen their ability to rebound from exhaustion. Aerobic fitness increases work capacity and the ability to withstand stress. While feelings of depression and moodiness accompany tiredness, aerobically fit Service members are affected less than those unfit. The ability to quickly recover from physically strenuous workloads is maintained by smart physical training, performed consistently and routinely. However, there is no evidence that good physical conditioning significantly reduces normal sleep requirements nor compensates for the deleterious impact of sleep deprivation on cognitive functioning. Sleep deprivation is discussed in greater detail in chapter 4.

2004. EFFECTIVE LEADERSHIP

The effective leader in combat is competent and reliable. He knows his job without question, and he can be counted on to do it regardless of the situation or circumstances.

Effective small-unit leadership reduces the impact of stress in several ways. Leaders understand the sources of combat stress and reactions to them. In addition, leaders manage stress problems to keep them from spreading throughout the organization by implementing the following actions:

- Continue mission performance; focus on immediate mission.
- Expect Service members to perform assigned duties.
- Remain calm, directive, and in control at all times.
- Let Service members know their reactions are normal and there is nothing seriously wrong with them.
- Keep Service members productive (when not resting) through recreational activities, equipment maintenance, and training to preserve perishable skills.
- Ensure Service members maintain good personal hygiene.
- Ensure Service members eat, drink, and sleep as soon as possible.
- Let the Service members express their thoughts. Do not ignore or make light of expressions of grief or worry. Give practical advice and put emotions into perspective.

A battalion commander in World War II made the following address to his Marines just before a new offensive:

> I know as well as you do that the going has been tough. Maybe I sweat it more than you do because I have more to worry about. But you've done a fine job, and I'm proud of every single man in my outfit. I assure you that everything possible will be done to give you the best available support, and I will not order you to attack unless I'm confident that you have a real chance to succeed . . . . The harder we fight now, the sooner we can finish this mess and get back to living the way we want.

That speech demonstrated a common sense approach in dealing with stress. Had this commander preached about the advantages of democracy over fascism or the evils of Hitler, his words would have fallen on deaf ears. Instead, he seized on the strong motivators of his troops: their *respect* for him as a leader, their desire to continue to be worthy of his respect, and their desire to share in the unit's accomplishments.

Generally, confidence is based on the technical and tactical competence exhibited by leaders and Service members. Since confidence is one of the strongest defenses against stress, its development is a major goal for every military leader. To gain confidence, Service members must believe in themselves, their equipment, other unit members, and their training. Above all, they must believe in their leaders' competence. Each of these beliefs is instilled and reinforced at every opportunity.

The following comments were made by a Marine colonel reflecting on his experience as a company commander in Vietnam.

> A feeling of helplessness will overtake men when they realize that they have been put in an untenable position. When the situation is so bad that the men cannot fight back, they are susceptible to fear. The

situation can be critical, but as long as the men can fight back, fear normally will not overtake them. The solution to this problem is: don't put your men in untenable positions. Granted, in every battle some men in a unit will find themselves in a situation where they can't fight back. This is inevitable especially in an attack. Proper use of fire team, squad, platoon, and company formations, as well as intelligent use of terrain, minimizes the danger of a large portion of the unit being pinned down to the extent where the men cannot return fire. If you are tactically and technically proficient, use common sense, apply the principles of war, and employ the firepower available to your unit, untenable situations can be avoided. Simply stated, know your job!

Those comments provide an insight and sensitivity to problems related to combat stress through small-unit leadership. His instincts told him how important it was to exhibit solid leadership qualities in order to allow his troops to operate at maximum efficiency. A unit builds confidence, esprit, integrity and cohesion when the *leaders know their jobs*.

2005. COMBAT STRESS MANAGEMENT BUILDING BLOCKS

Confidence in Leaders

Leaders must demonstrate effective leadership to earn their subordinates' loyalty and trust. Leaders are responsible for—

- Committing the unit to missions commensurate with abilities.
- Planning operations carefully and thoroughly.
- Preparing the unit to accomplish the mission.
- Leading and guiding the unit to mission accomplishment.

- Showing consistent good leadership that convinces subordinates their leaders know best what should be done, how it should be done, who should do it, and how long the task should take. Authority accompanies leadership beyond the automatic authority given by military rank and position. Authority and respect are earned based on confidence in a leader's ability to guide the unit to success.

Confidence in Training

Training helps Service members develop the skills required to do their jobs. Confidence is the result of knowing they have received the best possible training for combat, and are fully prepared. This confidence results from the following:

- Realistic training that ends with successful mastery.
- Relevance of training to survival and success on the integrated battlefield.
- Refresher and cross training.
- Systematic individual and collective training.

Confidence in Unit

Each Service member in a unit needs to become confident of the other unit members' competence. Individuals must stay and train together to gain that personal trust. Unless absolutely necessary, teams should not be disbanded or scrambled. Subunits in the same larger unit should have the same Standing Operating Procedures and training standards, so members can fit in quickly if teams have to be cross-leveled or reorganized after casualties occur. Confidence in the unit leads to feelings of security, which in turn allows members to sleep and positively focus stress. In combat, the unit must receive each member's highest commitment to unit loyalty. Mission accomplishment is the unit's highest priority.

Confidence in Equipment

Service members who learn to operate and maintain assigned equipment develop confidence in their ability to employ it. This, in combination with an individual's belief in his personal capabilities, raises overall confidence in fighting ability.

Cohesion and Morale

Good unit cohesion and morale offset the negative effects of combat stress. The foundation for any stress reduction program includes trust and confidence in the following:

- Fellow Service members.
- Competence and fairness of the unit leaders.
- Unit's technical abilities and military power.
- Equipment.
- Personal combat ability.
- Sense of support from the civilian community.
- Personal spiritual well-being.

Physical Conditioning

A strong relationship exists between physical stamina and the ability to resist combat stress. Good physical conditioning has physical and psychological benefits. Rigorous physical conditioning helps protect against the stress of continuous operations. A regular program of physical fitness to increase aerobic endurance, muscular strength, and flexibility is essential to combat readiness. As physical conditioning improves, Service members feel better about themselves, have greater confidence in each other, and their stress is reduced.

Unit training includes regular physical conditioning. This increases the members' tolerance to all types of stressors. The program is geared to the unit's combat mission, and exercises are tailored to the environment where the unit operates. The pace, length, and types of runs, road marches, and other activities are commensurate with the unit's need. Light infantry units need more demanding, longer road marches than maintenance units. Activities also include team athletics, which capitalize on cohesion-building aspects as well as physical benefits. The benefits of such a program include developing endurance through aerobic exercises, enhancing strength through weight training, and deprivation/physical stress training.

Family Care

Service members entering combat with financial worries or family problems risk breaking down under the additional stress of combat. Even positive but unfinished changes on the home front, such as a recent marriage or parenthood, can distract the Service members' focus on combat missions with worries that they will not live to fulfill their new responsibilities at home.

Leaders must be aware of this risk and assist members in handling personal matters before deployment. Pre-deployment and post-deployment family briefings should be conducted, and programs established to assist families before, during, and after deployment. When Service members know their families are cared for, they are better able to focus on their combat duties.

Coping with Individual Stress

Stress pushes the body to its limits and causes tension; relaxation reverses this process. Coping with personal stress is essential. Stress-coping skills are incorporated into unit training early, and command emphasis is placed on practicing them. Service members receive a block of instruction on stress-coping techniques, then are given supervised time each duty day to practice them quietly. After 3 or 4 weeks of practice, most Service members relax easily and quickly, even under highly stressful conditions. They will be able to naturally control stomach fluttering, heart rate, blood pressure, and stress.

The stress-coping exercises include deep breathing, muscle relaxation, and cognitive exercises. Deep breathing is the simplest to learn and practice; the others require longer instruction and more practice time. On request, the Combat Stress Control (CSC) team will provide instructional materials and assistance.

Deep-Breathing Exercise

This consists of slow, deep inhaling, which expands the chest and abdomen, holding it for 2 to 5 seconds, then exhaling slowly and completely, which pushes out the used air. This can be done for five breaths as a quick, mind-clearing exercise, or continuously to promote sleep. Abdominal or diaphragmatic breathing (making the stomach move the air, rather than the upper chest) is especially effective for stress control and, with practice, can be done simultaneously with task that require full attention.

Muscle Relaxation Exercises

These are more complex. They generally consist of concentrating on various muscle groups and the tensing and relaxing of limbs to relax the entire body. Quick versions for use in action consist of tensing all muscles simultaneously, holding for 15 seconds or more, then letting them relax and "shaking out the tension." Deep relaxation versions start in the feet and work up (or start in the head and work down), body part by body part (muscle group by muscle group), tensing and then relaxing each in turn, while noticing how each part feels warm after it relaxes.

Cognitive Exercises

These consist of self suggestion (positive self-talk); imagery (imagine being fully immersed in a deeply relaxing setting); rehearsal (imagine performing the stressful or critical task under pressure and doing it perfectly); and meditation (clearing the mind of all other thoughts by focusing on every breath and silently repeating a single word or phrase). These techniques involve creating positive mental images that reduce the effects of stressful surroundings, redirecting mental focus and learning to detach from stress. Service members are encouraged to practice stress management techniques and discuss their use in combat.

CAUTION

It is important not to use deep relaxation techniques at times when you need to be alert to dangers in your surroundings. Practice the quick relaxation techniques so you can use them automatically without distraction from the mission.

2006. CRITICAL EVENT DEBRIEF AND UNIT LEADER'S AFTER ACTION REVIEW

NOTE: On the subjects of Critical Event Debrief/After-Action Review, U.S. Army personnel will follow the guidance provided in FM 22-51, *Leaders' Manual for Combat Stress Control*, chapter 6.

Critical Event Debrief

A Critical Event Debrief (CED) is a structured group process designed to mitigate the impact of a critical event and to accelerate normal recovery of those personnel involved. The CED is normally conducted by a team composed of trained members, (medical officers, chaplains, mental health professionals, trained unit members). A CED's main value is to quickly restore unit cohesion and readiness to return to action, through clarifying what actually happened and clearing up harmful misperceptions and misunderstandings. It may also reduce the possibility of long term distress through sharing and acceptance of thoughts, feelings, and reactions related to the critical event. Ideally, CEDs are conducted 24 to 72 hours after the occurrence of the event, away from the scene and separate from any operational debriefing. The typical CED lasts from 2 to 3 hours. Situations that warrant a CED include:

- Death of unit member
- Death or suffering of noncombatants (especially women and/ or children)
- Handling of the dead, management of carnage or the sight of devastation during disaster work
- Serious friendly fire incident

- Situation involving a serious error, injustice or atrocity
- Terrorist attack.

After Action Review

Although a CED would be warranted following such traumatic events as those listed, if access to CED trained professional team is not available, small-unit leaders can modify their After Action Review (AAR) to assist in identifying the level of stress reaction of unit members. The AAR or "hotwash" is a routine practice used by leaders to debrief post-mission operations. Accordingly, it provides a familiar, non-threatening forum for leaders to identify the levels of stress reaction experienced by the unit members using the information in paragraph 1004. The leader's AAR may be sufficient to restore unit readiness for further action. It may also identify personnel who need immediate or later referral for chaplain or mental health/CSC help for combat stress-related symptoms. In those instances when access to CED trained facilitators is not available, the unit leader should incorporate the following questions in the AAR:

Fact Phase
Ask unit members to describe the event from their individual perspectives. What was each member's specific role in the event?

Thought Phase
What were each member's first thoughts at the scene (or when the incident was first heard of)?

Reaction Phase
What was the worst thing about the event? What was thought and/or felt?

Symptom Phase
Describe probable thinking, physical, and emotional responses both at the scene and a few days afterward.

Training Phase
Relay information regarding stress reactions and what can be done about them. If prepared handouts are available, distribute them. Include points of contact within the unit.

Wrap-up Phrase
Reaffirm positive things. Summarize. Be available and accessible. Debriefing team/unit leaders together decide which individuals need more help or referral.

2007. EFFECTIVE COMBAT STRESS CONTROL PROGRAM

An effective stress management program starts with early planning, continues during employment, and extends beyond the return home. The CSC program is comprised of three phases: predeployment, deployment and combat, and post combat.

Phase 1: Predeployment

During pre-deployment, some stressors facing Service members include long working hours, preparation for training, fear of the future, family worries, and anxiety about the unit's readiness. Signs of poor coping include insomnia, increased use of alcohol, marital problems, increased bickering in the unit, irritability, and suicidal feelings. The most important preparatory steps to take in the predeployment phase are to:

- Conduct unit training and mission rehearsals.
- Prepare for changed sleep schedules and jetlag.

- Attend to task assignments and allocations.
- Conduct equipment and supply maintenance checks.
- Attend to personal and family matters. (Call the Family Service Center.)
- Integrate new members into the unit positively and actively.
- Welcome "significant others" (not just entitled beneficiaries) in the Family Support network information tree.
- Brief as much information about the operation as possible, consistent with operations security measures.
- Actively familiarize the unit members with the stressors and horrors they may encounter (for example, watch videotapes from network television showing suffering civilians or atrocities), but always followed immediately with what the unit will do in those situations.
- Arrange for pre-deployment training and education, especially for refresher training of stress reduction techniques from local mental health professionals, a Special Psychiatric Rapid Intervention Team (SPRINT) or CSC team, if available.

Unit Training

Because unit leaders have combatted the stressors associated with garrison living and peacetime training, they have learned to know their Service members and what affects their performance. All Service members should believe unit training has prepared them thoroughly for combat. Stress is reduced if the training improves belief in themselves, and the unit's ability to fight and win. They are assured training is designed and developed to meet combat requirements, provides the ability to fight successfully, and that supporting units receive the same realistic training.

Realistic mission rehearsal helps desensitize Service members against potential combat stressors. For example, wearing and realistically training in protective gear is important. By doing so in pre-deployment training, Service members become less distressed in combat. The unit should be exposed to extensive drills wearing protective gear during live-fire exercises before employment. Such training has the two-fold, stress-reduction benefit of building confidence and preparing Service members for combat.

Leaders who provide their Service members with advance knowledge about the enemy prepare them for the stress created during deployment and initial enemy encounter. It is important during such training to talk realistically about enemy strengths and weaknesses as well as those of their own units. While inspirational pep talks are also important at this time, they should not include biased, inaccurate information. Leaders earn trust and respect if their troops perceive them as accurate, dependable sources of information.

Physical Fitness Training

In the time-urgent work of preparing for deployment, leaders assure that physical fitness is sustained, either by the work involved or by selective unit physical training exercises. New unit members who are insufficiently fit receive special training.

Stress-Coping Skills Training

During preparation for deployment, the leaders direct the unit to practice stress coping and relaxation techniques, and are positive role models by demonstrating use of these techniques. If necessary, the chaplain and mental health professional personnel available to the units can provide additional training.

Sleep Discipline

Before deployment, unit leaders must consider fatigue and sleep loss occurring during combat. The enforcement of work and rest schedules begins early in pre-deployment training. During continuous operations, fatigue caused by lack of sleep is a major source of stress. Breaks in combat are irregular, infrequent, and unscheduled. Extended sleep is unlikely. *Sleep logistics* is emphasized such that sleep and rest are allocated or supplied like rations, water, equipment, and ammunition. Sleep discipline training addresses the following points:

- A unit-specific work-rest-sleep plan is developed and practiced.
- The unit leader or commander is included in the allocation of sleep and rest time, as lack of sleep will impair his judgment and decision-making skills as much as those of his subordinates.
- The plan allows Service members at least 5 hours of uninterrupted sleep, ideally between 2400 and 0600, every 24 hours. Persons receiving only 5 hours per 24 hours over a period of several days will accumulate a significant sleep debt.
- Sleep priority is assigned to those whose judgment and decision-making are critical to mission accomplishment.
- If received frequently, 1 hour of sleep or even 15-minute naps help, but "slow mental starting" upon wake up can result.
- Relaxation exercises complement sleep schedules. These exercises are used as an alternative to regular sleep or as an aid to help Service members rest under difficult circumstances.

If preparation is made for jetlag, its effects are minimal. The ideal method is to seclude units up to several weeks before deployment, gradually changing work, eat, and rest cycles to coincide with the destination's schedule. Although rapid deployment does not permit this method, there are ways to help make the transition to a new environment, such as changing to new sleep cycles collectively versus individually; sleeping up to 12 hours before operation start; ensuring leaders receive rest; and adjusting diet to meet energy needs. If the unit is divided into shifts, make sure that shift assignments are maintained from day to day, i.e., if a member is assigned to the 0600-1400 shift, keep him on that same shift each day.

Task Allocation and Management

Overloading Service members with tasks or responsibilities is another major source of stress. Allocating tasks fairly among available Service members improves unit effectiveness as well as decreases stress. Proper allocation of tasks include:

Selecting the Right Person for the Job

The right person is fitted to the right task according to the task requirement and the individual's talents, abilities, and training.

Duplicating Critical Tasks

Two Service members are assigned to a critical task requiring mental alertness and complete accuracy. They check each other's work by performing the same task independently.

Cross-Training

Each Service member is trained in a secondary duty position to ensure competently stepping into the position of another.

Developing Performance Supports

Develop standing operating procedures, checklists or other mental aids to simplify critical tasks during periods of low alertness.

Equipment Maintenance and Supply

During pre-deployment, the unit maintains its equipment and manages needed supplies. Once deployed for combat, Service members require confidence that supplies are ample and equipment is dependable. The following questions are important:

- Does the unit provide ample training in equipment maintenance and troubleshooting?
- Has the unit's equipment been field-tested under realistic conditions? For example, have Service members fired and cleaned their weapons while wearing full combat gear or protective clothing?
- Does the unit have sufficient ammunition, food, water, and other essential supplies?
- Does the unit have contingency plans for procuring and managing critical supplies if normal channels are disrupted?

Personal and Family Matters

Family stress adds to combat-imposed stress and causes distraction, interference with performance of essential duties, and a negative impact on stress-coping ability. This will result in the unit's inability to perform at peak. The unit should help the Service members resolve important family care matters before deployment and develop methods for helping families when Service members are deployed. Service members are encouraged to:

- Generate or update their wills.
- Finalize power of attorney for spouses.
- Update life insurance policies, including Servicemen's Group Life Insurance (SGLI).
- Ensure family automobiles are in good repair.

- Develop lists of telephone numbers of reliable points of contact for specific problems (mechanics, emergency transportation, babysitters, sources of emergency money, health care, etc.).
- Resolve major legal issues such as alimony payments, property settlements following divorces, and child support payments.

Role of Leaders

Unit commanders should:

- Brief families as a group before deployment or as soon as possible after deployment. Within the bounds of security limits, explain the mission's nature. Even if a mission is highly confidential, families benefit from such a meeting by being told of the support available to them while separated. They begin to solve problems and form support systems with other families. This includes an opportunity to discuss family questions and concerns. The Family Services Office, base mental health service or the chaplain's office assist in staging this briefing.
- Establish points of contact (e.g. Key Volunteer Network) to assist with family problems. These volunteers possess good working relations with the chaplain and mental health personnel to assist with the management of complex problems.
- Establish key volunteer communication and support networks. Commanding officers' spouses, or spouses of sergeants major are often good resources for developing and running such networks; however, the involvement of junior Service members' spouses is also crucial. Some of the most enthusiastic participants are tasked to make *outreach* visits and encourage shy or depressed spouses to participate.
- Have mental health professionals conduct meetings to discuss post-deployment problems. For example, some children have

difficulty adjusting to a parent's absence, and other family members may experience similar difficulties. Mental health professionals give families valuable information on these normal reactions and suggest ways to prepare for them.

Phase 2: Deployment and Combat

In addition to the normal stress associated with moving to a combat zone, Service members in this phase start worrying about their survival and performance under fire. Their thoughts become centered on fear of the unknown. Leaders should emphasize that stress under these circumstances and conditions is expected and is a natural reaction. This will help prevent "normal" stress reactions from escalating into extreme reactions.

Unit leaders should provide as much information as necessary to their survival and mission success, reinforce the stress control techniques learned during pre-deployment, and help their subordinates understand what happens to them when stressors occur.

Deployment Vehicle

The deployment vehicle—in most cases, an airplane—is a stressor by itself. If it is a commercial aircraft, in-flight problems are usually minor. However, if the unit deploys on a military aircraft, leaders should accomplish the following:

- Designate areas for light exercise and stretching to counter seating discomfort caused by a buildup of leg fluids.
- Ensure Service members drink enough fluids to prevent dehydration and have access to head/latrine.
- Adopt the activity schedule of the new time zone. If the unit is in the sleep cycle or is already in or about to enter the sleep

cycle, cover windows; reduce lighting; and issue earplugs, blankets, and pillows.

- Allow uninterrupted sleep. If a stopover occurs during a sleep cycle, do not waken Service members to eat or partake in activities. If the stopover occurs during an activity period, take full advantage of it by having Service members take wash-cloth baths, stretch, and perform head-and-shoulder rotations.
- Upon arrival in the area of operations, follow the schedule of the new time zone. Eat the next meal and go to bed on the new schedule. Doing so helps the Service members' bodies adjust.

Information Flow

Since uncertainty about the future is a major source of stress, timely and accurate information becomes vital. Lines of communications are clearly defined and kept open. Issuing warning, operation, and fragmentary orders is critical to ensuring adequate information flow. Informational meetings are conducted at regular intervals, even when there is no new information to disseminate. This reinforces the organizational structure and the importance of unit meetings as the source of current, accurate information. Reliable sources of information are especially important for countering rumors.

Service members also need information or *performance feedback* after mission completion. Merely engaging in a firefight or completing a mission is insufficient. Service members must be told how they performed as a group. The knowledge of mission accomplishment and progress builds unit cohesion, develops a winning attitude, and reduces the effects of stress.

Family Support

Marine Corps Family Team Building programs and corresponding programs for other Services provide family support throughout deployments. The Navy-Marine Corps Relief Society, Army Emergency Relief, American Red Cross, Family Service Centers, and other community agencies also provide direct assistance to family members. The Key Volunteer Network and the American Red Cross continue to function as conduits for emergency information between Service members and their families. Leaders need to educate Service members about these programs and agencies that are available to serve the needs of the community. Effective communication and caring support networks help to prevent anxiety while Service members are deployed and/or in combat.

Religious Ministry Support and Pastoral Care

The Religious Ministry Team is an integrated part of the battalion or unit. In the Marine Corps, a Religious Ministry Team (RMT) consists of a chaplain and Religious Program Specialist (RP) or chaplain's assistant. RMTs are organic to commands or are assigned by higher headquarters to provide direct religious ministry support to a command. (**NOTE:** In the Army, the same teams are called Unit Ministry Teams (UMT).) During operations, the RMT often travels with the forward battalion aid station (BAS) in order to minister to the wounded and to be with the Service members who are most likely to experience combat stress. Using their professional training, skills, knowledge, and relationship with the Service members, chaplains provide care focusing on prevention of mild and moderate combat stress reactions. The RMT's primary mission is to provide ministry and pastoral care to the troops, offering faith, assurance, and hope. Before, during, and after the mission, the team provides field services, sacraments, and counsel, which bring comfort, assurance, and encouragement to Service members as they integrate their experiences into their lives.

In addition to being a spiritual/pastoral mentor for Service members, most chaplains are trained in some form of CED process. Chaplains are effective CED team members as well as trainers of small-unit leaders (e.g., platoon leaders, noncommissioned officers (NCOs), corpsmen, etc.) in CED team member skills and stress management techniques. Appendix B explains the roles of the Religious Ministry Teams.

Physical and Recreational Activities

Service members need an outlet for the anger, frustration, hostility, and grief developed in combat. It is unwise to conduct sports and recreation activities under observation by the enemy, but these activities can be conducted further to the rear. Although makeshift athletic games help relieve stress, they should be conducted with caution. It is interesting to note that the most frequent medical problem in Operation Desert Shield/Desert Storm was sports injury. Other activities, such as listening to music, reading or practicing relaxation exercises are encouraged at rest and recuperation facilities. Time and access to personal hygiene items and facilities are psychologically valuable in combat.

Humanitarian assistance and community relations' projects provide excellent opportunities for Service members to constructively engage in meaningful work. Public Affairs Officers, Civil Affairs Officers, and chaplains are excellent personal resources for developing working relationships with local nongovernmental organizations (NGOs) and private voluntary organizations (PVOs) in the coordination of such projects.

Integration of New Unit Members

The arrival of replacements requires small-unit leaders to conduct continuous training programs. Leaders should be concerned with the various stressors affecting new Service members, as well as those Service members who have been around for a while, including seasoned combat veterans. Leaders teach newcomers to use stress coping skills. Replacements are quickly integrated into their units and become thoroughly acquainted with all aspects of the new unit. The unit also becomes thoroughly acquainted with the replacements. A brief orientation with a sincere welcome, with NCOs and officer support, eases replacement transition for combat.

It is important Service members adjust to a new unit quickly and effectively. If they do not feel they are part of the group, unit cohesion and morale suffer, decreasing unit effectiveness. New Service members in garrison are assigned sponsors to assist them, introduce them to the unit, and smooth their families' adjustment. In combat, replacements are placed with suitable groups of seasoned veterans. New unit members are much more likely to become battle or stress casualties than are members of a seasoned and cohesive group. The veterans need to give support and advice to the newcomers, by example and direct action, if necessary.

Phase 3: Post-Combat

Just as pre-deployment and combat are stressful, the period after combat is also difficult. Today's rapid transportation enables Service members to travel from the battlefield to their hometowns in 48 to 72 hours. This short time often does not give them reflection with their comrades. Units should therefore set aside time in the last few days before leaving the theater to conduct their own "End of Tour" debriefing in which they start at pre-deployment and talk about whatever stands out in their memories, good or bad, as they recount the operation up to its end. There should also be appropriate memorial ceremonies and rituals that formally bring the operation to a close. Awards, decorations, and other recognition must be allotted fairly by the commanders.

Unit officers, staff NCOs, and NCOs, assisted by the chaplains and mental health/CSC teams, prepare the Service members for problems encountered during family reunion. For example, most Service members expect to resume roles and responsibilities they had prior to separation. However, their spouses often resist giving up their new roles as decision makers and primary home managers. Also, a spouse may feel that his or her sacrifices during the Service member's absence have gone unrecognized. This feeling becomes an additional source of tension. If at all possible, the families should receive the same briefings or written materials. Families need to be reassured of their contribution. Key Volunteer Networks and other Marine Corps Family Team building programs, and corresponding organizations for other Services continue to help manage problems with reunion and adjustment.

Service members are briefed that startle reactions to sudden noise or movement, combat dreams and nightmares and occasional problems with sleeping, and feeling bored, frustrated and out of place are common when first returning from combat to a peacetime, civilian setting. The leaders, chaplains, and the CSC team emphasize the normalcy of such reactions. Service members are also advised on resources available to help deal with such symptoms, if they are persistent and become upsetting.

Chapter 3

Command Leadership Actions

3001. LEADERSHIP ACTIONS AND INTERVENTIONS FOR COMBAT STRESS

It is important for military leaders to know something about the treatment of combat stress reactions. As most cases of initial combat stress symptoms do not require medical treatment, it has been found that military leaders are often quite adept at treating less severe cases of combat stress. In most cases, debriefing the mission, including any traumatic events, coupled with rest, food, and sleep, is often sufficient to alleviate the symptoms. If the operational tempo permits the Service member to remain with his unit, and he responds to simple reassurance (e.g., “You just need rest, you’ll be okay tomorrow”), he is not a casualty (by definition), and may not require further referral for specialized care. Military leaders in combat often make such decisions. When a Service member requires medical attention to rule out a possible serious physical cause for his symptoms, or because his inability to function endangers himself, the unit, and the mission, he should be transported to the BAS or equivalent nearest medical support facility. Interventions at the small-unit level include:

- If a Service member’s behavior endangers the mission, himself, or others, the leader should take appropriate measures to control him.
- If a Service member is upset, let him talk about what is upsetting him, listen, then try to reassure him.
- If a Service member’s reliability becomes questionable:

- Unload the Service member's weapon.
- Remove the weapon if there is a serious concern.
- Physically restrain the Service member only when safety is a concern or during transport.

- Reassure unit members that the signs are probably a normal combat stress reaction, and will quickly improve.
- If the combat stress reaction signs continue:
  - Get the Service member to a safer place.
  - Do not leave the Service member alone. Keep someone he knows with him.
  - Notify the senior NCO or officer.
  - Have the Service member examined by medical personel.
- If the tactical situation permits, give the Service member simple tasks to do when not sleeping, eating or resting.
- Assure the Service member that he will return to full duty as soon as possible.

The following quote was taken from an article about fear. It was written by a Marine who articulated his experiences during his tours in Vietnam. It provides intricate insight into how many Service members think and bond during combat.

> What should be done when men succumb to fear brought on by the realization that they are going to die? Before discussing this, first let us look at the men themselves. There are very few men who can be classified as cowards. Most men have too much self-respect to let their buddies down. It is the rare man indeed who will willingly violate the trust of his peers. The vast majority of men will give their lives rather than violate this trust. Proper training of combat troops, prudent leaders who are technically and tactically sound, and the reluctance of men to violate the trust of their peers are the foundation

of a solid combat unit. In this environment, courage and sacrifice are the rule, not the exception. At any given moment, anyone can be rendered ineffective by fear if one realizes that he is going to die. When this happens to a man, do not overreact. He knows he has let his buddies down, and he knows that his buddies and leaders know he has let them down. This is a very uncomfortable feeling for a combat Marine. The men around him will not make a big deal about his actions because they understand the situation. To make an issue of this situation will destroy the man and usually alienate the rest of your men. The man will most likely bounce back to his normal performance. The leader must keep in mind that today's coward is tomorrow's hero. If a man does not bounce back and continues to succumb to this type of fear, the Marines in his fire team will let you know when they have given up on him and no longer consider him trustworthy. Situations such as this are rare, so there is no need to make an example of the man. The biggest pitfall to be avoided is to judge the man's actions as a failure of your leadership as this will indeed cause you to overreact.

3002. SAFETY CONSIDERATIONS

Confiscation of a Service member's weapon should only be considered when it is clearly apparent that the Service member is unreliable and a safety hazard to himself or others. Service members identify with other Service members who are trusted and needed by their comrades. This is one of the strongest factors that aid in returning battle-fatigued members to effective duty. The small-unit leader relocates battle-fatigued Service members to a safer, quieter place to rest and work for several hours up to 1 or 2 days. If the unit cannot wait for the Service members to recover, it is necessary to evacuate them to the first level supporting medical unit. From there, every effort is made to move the Service members to a nonmedical unit or area (a tent or building of opportunity could suffice) for rest, replenishment, and reassurance. It should be made clear that the Service members are tired and in need of an opportunity to talk, to sleep, eat, and replenish fluids; they are not patients. Some small detachments without first sergeants or platoon/company organizations are attached to other units for support and rely on the effectiveness of small-unit leaders. Each Service member is accounted for and every effort is made to ensure strong lines of communications are in place and maintained between Service members and their original unit. Key to successful treatment is the return of the Service member to his original unit.

Actions to be taken for severely combat-stressed Service members are the same as those for the moderately combat-stressed, with one exception: medical personnel at the BAS level should evaluate severely combat-stressed Service members as soon as possible. Casualties will be treated and released within hours; held for rest and replenishment; or evacuated for further restoration. Service members who recover from combat stress reactions; return to their original units, i.e., same company or platoon; and are welcomed upon their return are less likely to suffer recurrence. Once rested and returned, they usually become healthy again. New joins and replacements in the unit are more susceptible to succumb to combat stress. Accordingly, this risk is reduced when Service members recovering from combat stress return to the same unit where their combat experience is known and welcomed. In rare instances, however, it is in the best interest of the individual to be reassigned to other jobs or units.

3003. COMBAT STRESS CASUALTY INTERVENTION MODELS

Management of combat stress reactions is unlike the treatment of physical trauma. Severely injured Service members are stabilized as rapidly as possible and then transferred to the rear. In all wars since World War I, combat stress casualties treated in the rear rarely return to full duty. In contrast, when the same casualties are treated near the front, approximately 75 percent return to full duty. Of those returned to full duty, only 10 percent experience continued symptoms requiring further treatment. Some studies suggest half of those treated at the rear go on to have chronic psychiatric symptoms, and approximately half return to full duty. Therefore, it is clearly in the Service member's best interest to be treated at the front and returned to duty. This also serves the best interest of the unit because it receives a veteran Service member back to duty, rather than breaking in a replacement.

Guidelines for dealing with Service members unable to function because of combat stress reaction are summarized in the memory aid **BICEPS.** BICEPS stands for brevity, immediacy, centrality (Marines)/command contact (Army), expectancy, proximity, and simplicity.

- **Brevity.** Critical Event Debriefing should take 2 to 3 hours. Initial rest and replenishment at medical CSC facilities should last no more than 3 or 4 days. Those requiring further treatment are moved to the next level of care. Since many require no further treatment, military commanders expect their Service members to return to duty rapidly.
- **Immediacy.** CSC should be done as soon as possible when operations permit. Intervention is provided as soon as symptoms appear.

- **Centrality (Marines)/Contact (Army).** Service members requiring observation or care beyond the unit level are evacuated to facilities in close proximity to, but separate from the medical or surgical patients at the BAS, surgical support company in a central location (Marines) or forward support/division support or area support medical companies (Army) nearest the service members' unit. It is best to send Service members who cannot continue their mission and require more extensive respite to a central facility other than a hospital, unless no other alternative is possible. The Service member must be encouraged to continue to think of himself as a warfighter, rather than a *patient* or a *sick person*. The chain of command remains directly involved in the Service member's recovery and return to duty. The CSC team coordinates with the unit's leaders to learn whether the overstressed individual was a good performer prior to the combat stress reaction, or whether he was always a marginal or problem performer whom the team would rather see replaced than returned. Whenever possible, representatives of the unit, or messages from the unit, tell the casualty that he is needed and wanted back. The CSC team coordinates with the unit leaders, through unit medical personnel or chaplains, any special advice on how to assure quick reintegration when the Service member returns to his unit.

- **Expectancy.** The individual is explicitly told that he is reacting normally to extreme stress and is expected to recover and return to full duty in a few hours or days. A military leader is extremely effective in this area of treatment. Of all the things said to a Service member suffering from combat stress, the words of his small-unit leader have the greatest impact due to the positive bonding process that occurs during combat. Simple statements from the small-unit leader to the Service member that he is reacting normally to combat stress and is expected back soon have positive impact. Small-unit leaders should tell Service members that their comrades need and expect them to return. When they do return, the unit treats them as every other Service member and expects them to perform well. Service members suffering and recovering from combat stress disorder are no more likely to become overloaded again than are those who have not yet been overloaded. In fact, they are less likely to become overloaded than inexperienced replacements.

- **Proximity.** In mobile war requiring rapid and frequent movement, treatment of many combat stress cases takes place at various battalion or regimental headquarters or logistical units, on light duty, rather than in medical units, whenever possible. This is a key factor and another area where the small-unit leader helps in the treatment. CSC and follow-up care for combat stress casualties are held as close as possible to and maintain close association with the member's unit, and are an integral part of the entire healing process. A visit from a member of the individual's unit during restoration is very effective in keeping a bond with the organization. A Service member suffering from combat stress reaction is having a crisis, and there are two basic elements to that crisis working in opposite directions. On the one hand, the Service member is driven by a strong desire to seek safety and to get out of an intolerable environment. On the other hand, the Service member does not want to let his comrades down. He wants to return to his unit. If a Service member starts to lose contact with his unit when he enters treatment, the impulse to get out of the war and return to safety takes over. He feels that he has failed his comrades and they have already rejected him as unworthy. The potential is for the Service member to become more and more emotionally invested in keeping his symptoms so he can stay in a safe environment. Much of this is done outside the Service member's conscious awareness, but the result is the same. The more out of touch the Service member is with his unit, the less likely he will recover. He is more likely to develop a chronic psychiatric illness and get evacuated from the war. This is one of the essential principles of CSC.

- **Simplicity.** Treatment is kept very simple. CSC is not therapy. Psychotherapy is not done. The goal is to rapidly restore the Service member's coping skills so that he functions and returns to duty again. Sleep, food, water, hygiene, encouragement, work details, and confidence-restoring talk are often all that is needed to restore a Service member to full operational readiness. This can be done in units in reserve positions, logistical units or at medical companies. Every effort is made to reinforce Service members' identity. They are required to wear their uniforms and to keep their helmets, equipment, chemical protective gear, and flak jackets with them. When possible, they are allowed to keep their weapons after the weapons have been cleared. They may serve on guard duty or as members of a standby quick reaction force.

If treated near their units, 65 to 85 percent of combat stress casualties treated in medical CSC facilities return to duty within 1 to 3 days. About 15 to 20 percent more return to duty in 1 to 2 weeks. Only 5 to 10 percent are sent home, and these usually have other problems in addition to combat stress reactions. If evacuated, few combat stressed Service members will return to duty. In fact, many are likely to be permanently disabled.

History shows that it is important to treat combat stress casualties as close to the front as practicable. Everyone is susceptible to the effects of combat stress, regardless of past performance, rank or duty. *Most Service members receiving combat stress control in accordance with the principles of BICEPS return to duty quickly.* The chances of later psychiatric problems are also reduced.

Combat stress reactions are inevitable in combat, but high stress casualties are not. History shows that highly trained and small cohesive units with good leadership have less than 1 such casualty for every 10 to 15 WIA, even in very heavy fighting. This is significantly fewer than the usual 1:3 to 1:5 ratio in high-intensity battle, and the 1:1 or 1:2 ratio in extreme fighting. Knowledge of factors that increase combat stress disorder in the tactical and overall situation encourages small-unit leaders, comrades, and the individual Service member to act towards building unit cohesion, sharing the burden, and reducing stress. The same measures, plus positive adherence to discipline and the law of war, also prevent combat stress-related misconduct that defeats the purpose of the mission. Tough, realistic training that builds confidence and caring within the unit in combat overcomes many adversities associated with combat stress in a combat environment. However, if a member fails to adequately respond to actions taken by the unit, then he should be evacuated for appropriate medical evaluation.

Chapter 4

Sleep Deprivation

4001. CHALLENGES OF SLEEP DEPRIVATION

People accumulate a "sleep debt" (cumulative loss of sleep over time) when they perform under limited sleep conditions. The only corrective measure for satisfying this sleep debt is sleep itself. Military operations, by their demanding nature, create situations where obtaining needed sleep will be difficult or impossible for more than short periods.

Continuous operations are military operations with many pulses of action every day and night, continuing for several days to weeks, which require careful planning and resource allocation to give everyone a minimum of 4 hours sleep in 24. (FM 22-51)

Sustained operations are continuous operations or combat with opportunity for less than 4 hours sleep per 24 hours for significant personnel, which may be brief or fragmented. (FM 22-51)

Accordingly, service members may have opportunities for only limited or fragmented sleep over an extended period. As a result of these periods of sleep loss, several combat tasks are likely to show decreased performance. These tasks include the following:

- Orientation with friendly and enemy forces (knowledge of the squad's location and maintaining camouflage, cover, and concealment).

- Coordination and information processing (coordinating firing with other vehicles and dismounted elements, reporting vehicle readiness, and communicating with the headquarters).
- Combat activity (firing from bounding vehicle, checking the condition of weapons, observing the terrain for enemy presence).
- Force preservation and regrouping (covering disengaging squads, marking the routes between locations, and conducting reconnaissance).
- Command and control activity (directing location repositioning, directing mounted defense, assigning fire zones and targets).

Continuous operations will potentially be more commonplace on the battlefield. In offensive operations, darkness is the time to retain or gain the initiative; while in defensive operations, obstacles can be employed with greater security during darkness. Forces can disengage undetected and threats to close air support lessen. The physical environment changes at night. As the air cools below ground temperature, inversions reduce visibility and hamper radar and radio signals. Conditions are optimal for using chemical weapons. Visual changes also occur. Without the aid of white light, there is no color perception. There is also a decrease in visual clarity, field of view, and depth perception. Targets take longer to engage. Preparation time increases two-fold to six-fold. Simple actions, such as the departure and return of patrols, become more complex and dangerous. Nighttime planning and coordination require greater attention. Navigation, adjusting fire, and munitions and/or target matching are more difficult. Precision is essential, but accuracy has a price. Service members tend to maintain accuracy at the sacrifice of speed. The adverse conditions associated with or generated by continuous ground combat at night will degrade the fighting performance of Service members, teams, and units. The almost complete mechanization of land combat forces and technological advances that permit effective movement at night, during poor weather conditions, and under conditions of limited visibility have largely overcome the reasons for "traditional" pauses in battle, such as darkness, resupply, and regrouping. New technologies have significantly increased the range, reduced the time, and changed the conditions over which battles are fought. For example, day/night-capable vehicles can operate for extended periods without re-supply, but they are limited by a crew's need to sleep. A Service member is not a machine and is, therefore, the weak link in the chain. The equipment can operate longer than the Service member who operates it, as the Service member must have sleep.

Commanders and leaders must ensure that all Service members obtain enough rest to counteract the effects of rapidly shifting from daytime to nighttime duty hours, or to extended work schedules. Implementing countermeasures that are designed to help Service members adapt to continuous operations conditions can satisfy this requirement. Neither leaders nor their subordinates can perform without rest or sleep. The Service member, the unit, and the leader are all affected by continuous operations. Generally at night, the cognitive and physiological resources of Service members are not at their peak, especially after a rapid shift from daytime to nighttime duty hours. Fatigue, fear, feelings of isolation, and loss of confidence may increase.

Non-stop, unrelieved combat operations (sustained operations) with little or no sleep degrade performance and erode mental abilities more rapidly than physical strength and endurance. Information gained from the Army Unit Resiliency Analysis Model shows that even healthy young Service members who eat and drink properly experience a 25 percent loss in mental performance for each successive 24-hour period without sleep. The mental parameters include decisionmaking, reasoning, memory tasks, and computational tasks. The loss may be greater for Service members who are older, less physically fit, or who do not eat and drink properly.

The effects of sustained operations are sometimes hidden and difficult to detect. Units are obviously impaired when Service members are killed or wounded in action or become noncombatant losses. They are further impaired when their troops are too tired to perform their tasks. Unlike individual performance, unit performance does not deteriorate gradually. Units fail catastrophically, with little warning.

A priority for fighting units is to assure that commanders and leaders are rested and able to think clearly. While this is obvious, it is a most difficult lesson for leaders to learn. During combat, commanders must focus on the human factor. They must assess and strengthen their units as they plan and fight battles. They must accurately decipher which units must lead, which must be replaced, where the effort must be reinforced, and where tenacity or audacity and subsequent success can be exploited. When leaders begin to fail, control and direction become ineffective, and the organization disintegrates. No fighting unit can endure when its primary objectives are no longer coordinated. Leaders must also prepare and precondition Service members to survive. It is particularly important that leaders conscientiously plan and implement effective sleep plans, because activities that are most dependent on reasoning, thinking, problem solving, and decision-making are those that suffer most when sleep and rest are neglected.

Some leaders wrongly believe that their round-the-clock presence during an operation is mandatory; they are unwilling to recognize that they, too, are subject to the effects of sleep deprivation. If the unit has been regularly trained according to the mission command philosophy, two benefits accrue. Not only will a leader be confident that in his absence his subordinates will adhere to his intent, but the trust he shows in his subordinates will continue to maintain unit morale and help ease some of the stress of the situation.

In future operations, the battlefield will become increasingly lethal. The threat of nuclear, biological, and chemical weapons will maximize confusion, uncertainty, and stress, which adversely impact our ability to move, shoot, communicate, and sustain. Sleep loss in this type of environment increases an already stressful situation.

4002. EFFECTS OF SUSTAINED OPERATIONS ON PERFORMANCE

A basic rule for continuous operations is planning ahead to avoid sustained operations, and provide members 5 to 6 hours sleep in 24. However, missions or enemy actions sometimes require exceptional exertion for several days with only unpredictable, fragmented sleep—as required in sustained operations. Sustained combat leads to exhaustion and reduction in effective task performance. Even during the first night of combat, normal sleeping habits and routines are abnormal. The Service member feels the effects of fatigue and the pressure of stress from noise, disrupted sleep time, and threat to life. While essential for endurance, sheer determination cannot offset the mounting effects of adverse conditions. Cognitive degradation involving poor decisionmaking begins during and after the first 24 hours of sleep deprivation.

Individual and unit military effectiveness is dependent upon initiative, motivation, physical strength, endurance, and the ability to think clearly, accurately, and quickly. The longer a Service member goes without sleep, the more his thinking slows and becomes confused. Lapses in attention occur, and speed is sacrificed to maintain accuracy. Continuous work declines more rapidly than intermittent work.

Tasks such as requesting fire, integrating range cards, establishing positions, and coordinating squad tactics become more difficult than well-practiced, routine physical tasks, such as loading magazines and marching. Without sleep, Service members can perform the simpler and/or clearer tasks—lifting, digging, and marching—longer than the more complicated or ambiguous tasks such as a fine hand-eye coordination sequence; i.e., tracking a target through a scope.

Sleep loss affects memory, reasoning, mental assessments, decision-making, problem-solving, subsequent actions, and overall effectiveness. While comprehension is accurate, reading speed slows and recall fails. For example, Service members may understand orders when reading them in documents, yet they are forgotten later when required. Individuals will forget or omit assigned tasks more often than they will make errors in carrying them out.

Leaders can expect declining moods, motivation, initiative, planning ability, and preventive maintenance. High motivation will only increase risk, due to impaired performance. Leaders must recognize erratic or unreliable task performance in subordinates, as well as in themselves. Alertness and performance decline gradually with partial sleep deprivation; that is, when sleep is limited to 4 to 5 hours per night. After 5 to 7 days of partial sleep deprivation, alertness and performance decline to the same low levels as those following 2 days of total sleep deprivation. After 48 to 72 hours without sleep, personnel become militarily ineffective.

Adverse Conditions

Continuous combat forces Service members to perform under adverse conditions that cause degradation in performance. Examples of adverse conditions follow.

Low Light Level

The amount of light available for seeing landmarks, targets, and maps is greatly reduced at twilight and night.

Limited Visibility

Smoke, fog, rain, snow, ice, and glare degrade a Service member's ability to see his environment and objects within it, as opposed to situations free of such conditions.

Disrupted Sleep Routines

People are accustomed to being awake or asleep during certain hours of the day or night. Disruption of the normal sleeping schedule causes degraded performance.

Physical Fatigue

Working the muscles faster than they can be supplied with oxygen and fuel rapidly creates "oxygen debt," eventually making these muscles unable to function until the deficits are made up during brief rests.

Sleep Loss

The muscles can continue to function adequately without sleep, but the brain cannot. Increasing sleep debt leads to subtle, but potentially critical, performance failures.

Sleep Loss Indicators

Indications of degraded performance symptoms become more prevalent as sleep debt accumulates. Performance is affected by the hours of wakefulness, tolerance to sleep loss, and the types of mental or physical work. Both mental and physical changes occur, with symptoms varying among individuals. Leaders must observe Service members for the following indications of sleep loss and degraded performance:

- Physical changes in appearance, including vacant stares, bloodshot eyes, pale skin, and poor personal hygiene. Other physical signs of sleep loss include the body swaying when standing, sudden dropping of the chin when sitting, occasional loss of hand-grip strength, walking into obstacles or ditches, low body temperature, slowed heart rate, and slurred speech.

- Mood changes, decreased willingness to work, and diminished performance go hand-in-hand. Service members may experience decreasing levels of energy, alertness, interest in their surroundings, and cheerfulness with a concurrent increase in irritability, negativity, and sleepiness. Some become depressed and apathetic. Others, for a time, can become energized by sleep loss, talk more, and may be more assertive without necessarily maintaining good judgment. Sleepiness and mood changes are not signs of weakness. After long periods of sleep loss, Service members go from being irritable and negative to dull and weary.

- Service members may feel more effort is needed to perform a physical task in the morning than in the afternoon. Exaggerated feelings of physical exertion may lead to work stoppage, especially between 0400 and 0700. During that time, the tendency to fall asleep is considerably more noticeable than other times.

- Both bickering and irritability increase with sleep loss. When Service members argue, it shows that they are still talking to each other and exchanging orders and messages. When arguments cease, especially after a period of increased bickering, Service members may be in a state of mental exhaustion.

- Comprehension and perception slow considerably. Individuals require extended time to understand oral, written or coded information; to find a location on a map and/or chart coordinates; to interpret changes in enemy fire patterns; and to make sense of things seen or heard, especially patterns. They may have difficulty with spot status or damage reports, and may be unable to assess simple tactical situations.

Loss of Concentration

Sleep deprivation causes the attention span to shorten. There is a loss of concentration on the job as dream-like thoughts cause lapses in attention. Leaders should watch for the following:

- Decreased vigilance. Personnel are less alert and fail to detect the appearance of targets, especially in monotonous environments. They may doze off at the wheel of moving vehicles.

- Distorted attention. Service members may imagine seeing things that are not there, e.g., "moving" bushes when in reality there is no such movement. The sleep-deprived brain can also misperceive bushes, rocks, people, vehicles or anything else and see them as something different, in very precise detail. Often the tired brain "sees" what it wishes were there (food, a bed); at other times, these illusions may be animals or other more bizarre things. But when the mind is alert for an enemy, the brain may generate a very convincing, detailed image of the enemy. Sometimes, but not usually, sounds or other sensations may accompany these illusions. They usually last only seconds, but can persist for minutes if not challenged, and rarely have even been "seen" by equally sleep-deprived comrades when told of them. It is essential for sleep-deprived unit members to check out any questionable things they see with their comrades, and to faithfully follow reporting and challenge procedures.

- Inability to concentrate; easily confused. Service members cannot keep their minds on what they are doing. They cannot follow multiple directions nor perform numerical calculations.

- Failure to complete routine tasks. Sleep loss interferes with completing routine individual tasks, such as drying the feet, changing socks or filling canteens when water is available. Tasks such as performing weapons checks may be skipped.

When a Service member cannot recall what he just saw, read, heard or was told by another individual, he is exhibiting a common sign of sleep loss. His memory loss is limited to recent events. For example, a sleep-deprived Service member may forget recent target data elements or recall them incorrectly and have difficulty learning new information.

4003. ACHIEVING SLEEP IN COMBAT

Sleep deprivation produces stress and, therefore, sleep management is important. Sleep management is a combat multiplier. Planned sleep routines are important for keeping the unit, the individual Service members, and the leader himself functioning as required while reducing sleepiness during continuous combat. Since leaders are responsible for planning sleep routines, they need a basic understanding of the physiological and behavioral aspects of sleep and their impact on performance. The following paragraphs provide this information.

Rhythmic Variations

There are rhythmic variations in individual performance based on a predictable physiological and behavioral cycle that comprises about 24 hours. The 24-hour, day-night/work-rest cycle is called the *circadian rhythm*. Because traveling across a half-dozen time zones disrupts the usual relationship in the day-night/work-rest cycle, for a few days Service members are not sleepiest at their usual sleep period of 2400 to 0600, new-locale time. Allowing sleep about 1200 to 1800, new-locale time, will only delay their adaptation to their new locale. Leaders must instruct troops to go to bed between 2400 and 0600 new-local time to establish a new circadian rhythm.

Another example of circadian rhythm is body temperature. Although one's "normal" temperature is 98.6 degrees, this is really an average or midpoint of a daily swing from 96.8 to 100.8 degrees. For someone accustomed to working days and sleeping nights, body temperature would fluctuate approximately as indicated. There is a well-established link between body temperature and sleepiness and/or performance slumps. Performance parallels body temperature. The higher the body temperature, the better the performance. As body temperature decreases, mood and motivation decline with a concurrent increase in sleepiness and fatigue.

Impact upon performance is most pronounced during the *circadian lull*, which is roughly 0200 to 0600 hours. During this time, performance declines about 10 to 15 percent. In sleep-deprived Service members, this decline may reach 35 to 40 percent. If the day-night/work-rest cycle is disrupted, performance suffers because the Service member is sleepy during the new work period and awake during the new sleep period. The body needs several days to adjust to the new schedule. Critical hours for sleep are between 0200 and 0600 when *anchor sleep* (the most beneficial sleep) is taken. The body is at its lowest temperature during this period. This is the best time for sleeping, but not for napping. To prevent sleep inertia, naps should always be taken at times other than the lowest point in body temperature.

Leaders need to calculate the difference in time zones and make the necessary schedule changes. Leaders will need day-and night-fighting teams. Members acclimated to working days and sleeping nights should be scheduled to work nights and sleep days.

Their performance slump/optimal time to sleep would be 2400 to 0600, new-locale time. Deployment, pre-combat, and combat are not usual circumstances. If certain Service members must have an offset circadian timing from the rest of the unit, a special effort must be made to establish their sleeping time. Obviously, troops must sleep whenever possible. If a planned sleep schedule cannot be followed, however, performance is enhanced if sleep coincides with the low point in body temperature.

Adjusting to new circadian rhythms is a slow process, taking 3 to 6 days to come "in phase" with a new schedule. Leaders should devise a sleep schedule that provides for sleep at the same time of day or night every 24 hours. Sleep schedules that provide for sleep at different times of day or night are less valuable and are detrimental to quality sleep and optimal performance.

Sleep Shifts

Staggered work schedules can be set up for two shifts working 4 hours on/4 hours off, 6 hours on/6 hours off, and 12 hours on/12 hours off. See Table 4-1. Each shift follows the same schedule daily. It is better to maintain regular shift schedules than schedules that continually change.

Sleep/Rest Guidelines

Leaders should use the following sleep and/or rest guidelines in this section to enhance individual and the unit performance in continuous operations.

- Know personal tolerance for sleep loss and those under your command; major individual differences are not easily changed. Individuals who are unable to sleep during pre-deployment and deployment stages should be encouraged to practice relaxation exercises (see paragraph 2005).

Table 4-1. Sleep Shifts.

4 HOURS ON/4 HOURS OFF						
Shift	2400-0400	0400-0800	0800-1200	1200-1600	1600-2000	2000-2400
1	SLEEP	DUTY	SLEEP	DUTY	SLEEP	DUTY
2	DUTY	SLEEP	DUTY	SLEEP	DUTY	SLEEP

6 HOURS ON/6 HOURS OFF				
Shift	2400-0600	0600-1200	1200-1800	1800-2400
1	SLEEP	DUTY	SLEEP	DUTY
2	DUTY	SLEEP	DUTY	SLEEP

12 HOURS ON/12 HOURS OFF		
Shift	2400-1200	1200-2400
1	SLEEP	DUTY
2	DUTY	SLEEP

- Ensure that Service members fully use their breaks and other opportunities for rest. Encourage them to waste no time in getting to sleep. Undisturbed, prolonged sleep is the most desirable use of rest opportunities. When there has been sleep loss but little physical exertion (e.g., manning communications, operating a radio), mild physical exercise such as walking around when conditions permit, can help maintain alertness.
- Encourage Service members to sleep, not just rest, by creating the most conducive environment possible for sleep: quiet, without interruptions (or earplugs); dimness or darkness (or with eye cover); not overly warm or cold.

- Do not allow personnel to sleep in unsafe conditions. Enforce strict rules designating sleep areas and requiring perimeter guards. Require day and night guides for all vehicles to prevent Service members from being accidentally run over.
- Ensure that Service members follow sleep schedules or routines. The field commander who does not enforce a sleep schedule or routine leads his troops into an environment that increases the opportunity for hazardous conditions to be encountered while in continuous combat. Taking naps is not a sign of low fighting spirit or weakness; it is a sign of foresight.

Measuring Sleep Loss

Sleep loss can be measured by:

- Keeping a sleep and/or activity log. From pre-deployment to post-deployment, log sleep and nap periods. Service members need 4 to 5 hours per 24-hour period; 6 or 7 hours is optimum. If they receive less, the first chance for a long rest period must be used for sleep.
- Observing performance and asking questions. Look for the indications of sleep loss—such as increase in error occurrence, irritability, difficulty understanding information, and attention lapses—with concurrent decreases in initiative, short-term memory, and attention to personal hygiene. Confirm sleep loss by asking the obvious question: "When did you sleep last and how long did you sleep?"

Sleep Loss Alternatives

Ways to overcome performance degradation include:

- Upon signs of diminished performance, find time for members to nap, change routines or rotate jobs (if cross-trained).

- Have the Service members most affected by sleep loss execute a self-paced task.
- Have Service members execute a task as a team, using the buddy system.
- Do not allow Service members to be awakened for meals while in flight to a new location, especially if the time zone of the destination is several hours different than that of point of departure.
- Insist that Service members empty their bladder before going to bed. Awakening to urinate interrupts sleep, and getting in and out of bed may disturb others and interrupt their sleep.
- Allocate sleep by priority. Leaders, on whose decisions mission success and unit survival depend, must get the highest priority and largest allocation of sleep. Second priority is given to Service members that have guard duty and to those whose jobs require them to perform calculations, make judgments, sustain attention, evaluate information, and perform tasks that require a degree of precision and alertness.

4004. SLEEP/REST PLANNING

Sleep/rest planning applies to the pre-deployment, deployment, pre-combat, combat, and post-combat stages of battle.

Pre-Deployment Stage

Using mission-scenario operation guidelines, determine periods available for sleep and the total number of sleep hours possible. Because continuous operations requirements may change, alternate sleep routines should be planned. Become familiar with the area where the combat unit will sleep; For example, some may have to sleep in mission-oriented protective posture (MOPP) IV. If sleeping in MOPP IV is anticipated in combat, practice it during the pre-deployment stage. Prior experience reduces stress, so practice anticipated sleep routines before continuous operations.

Deployment Stage

Since sleep will be reduced during deployment, follow preplanned sleep routines. The prudent commander will choose a 4-hour on/4-hour off, 6-hour on/6-hour off, or 12-hour on/12-hour off shifts from the start. Take into account that Service members on night duty will need to sleep during the daytime. Provide night-shift personnel with separate sleeping quarters to avoid disruption of their sleep period.

Pre-Combat Stage

In general, people are most effective during the afternoon and are least effective just before dawn. Without prior adjustment to the new time zone, which naturally occurs in 3 to 5 days, leaders can expect degraded daytime performance. The reason is that 0200 to 0600 hours home-base time is the low point in performance efficiency and should be considered when planning workloads.

Combat Stage

Every effort should be made to avoid situations where all personnel are physically and mentally exhausted simultaneously. Make the most of any lull during the combat phase by sleeping briefly. Complete recovery from sleep loss may not be possible during intense combat, but limited sleep is helpful. Uninterrupted short sleeps of 15 minutes or longer are beneficial to partially recovering alertness. Sleep during the combat stage may be risky, however, because a Service member may wake up feeling groggy, confused, sluggish, and uncoordinated. It may take his brain from several seconds to 15 minutes to "warm up." Individuals differ in how quickly they take to wake up, but it tends to be worse when the body expected to go into deep sleep, and to get worse with increasing sleep loss. Activities that increase circulation of warm blood to the brain, like moderate exercise or drinking a hot beverage, may shorten the start-up time.

Post-Combat Stage

It is important to make up sleep debt, but experts disagree about the amount of recovery time needed. Some say the hours of sleep needed for recovery after sleep deprivation are less than the amount lost. It is well known and documented that lost sleep is not made up hour-for-hour. Most experts agree that immediately following continuous combat, Service members should be allowed to sleep up to 10 hours. Longer sleep periods are not desirable because they cause "sleep drunkenness" and delay in getting back to a normal schedule. After the first sleep period of up to 10 hours, Service members should return to the regular sleep routine. Sleep inertia lasting longer than 5 to 15 minutes and increased sleepiness may occur for as long as a week following sustained combat. Some experts recommend that 4 of the first 8 hours of recovery sleep should be at the 0200 to 0600 sleep time, and they suggest the following guidelines for complete recovery from the effects of sleep loss:

- 12 hours for sleep and rest after 36 to 48 hours of complete sleep loss with light to moderate work load (fatigue may linger for 3 days).
- 24 hours for sleep and rest after 36 to 48 hours of sleep loss with high workload (12 to 16 hours per day).

- 2 to 3 days time off after 72 hours or more of acute sleep loss.
- As much as 5 days for sleep and rest following 96 hours or more of complete sleep loss.

Most experts agree that 10 hours of sleep is the maximum needed, with the additional 2 hours used for rest. It is doubtful that a Service member could continue past 72 hours of wakefulness. Should this occur, a couple of nights with 10 hours of sleep are more beneficial than an excess of 10 hours during one sleep period. If Service members have not slept for 36 to 48 hours or more, they should avoid sleep of less than 2 hours, especially between 0400 and 0600. A too-short sleep period at the wrong time may cause a long period of sleep inertia. After 96 hours of total wakefulness, 4 hours of sleep may provide substantial recovery for the simpler, less-vulnerable tasks. Recovery continues with additional days of 4 hours of sleep per 24 hours. Complex leadership tasks may require longer recovery sleep, but sleep until fully satisfied is not necessary.

Sleep loss alone does not cause permanent health problems, nor does it cause mentally healthy people to become mentally ill. Reduced sleep (from 8 to 4 hours) does not cause physical harm. Hallucinations may occur, but they disappear after recovery sleep. Clinical laboratory tests show that total sleep loss of over a week does not pose serious health problems. It is doubtful that Service members could stay awake for such an extended period, and it is not suggested that Service members try to endure long periods without rest. However, the effects of sleep loss, such as inattentiveness and poor judgment, may be harmful (such as falling asleep at the wheel of a vehicle).

Sleep cannot be stored in our bodies for emergency use. Sleep of more than 7 to 8 hours before deployment does not "store up" excess sleep, but sleep taken immediately before a deployment can prolong activity. Therefore, it is important to begin continuous operations in a rested state. During daytime or early morning naps, many Service members experience vivid dreams as they fall asleep and often wake up frightened. Leaders should inform their troops that this occurrence is both common and normal during daytime sleep. If a single, unbroken period of 4 to 5 hours is not available for sleep, "power naps" of 15 to 30 minutes, although less recuperative, can be taken. Leaders must capitalize on every opportunity for a "power nap." Merely resting by stretching out does not take the place of sleep. Only sleep can satisfy the need for sleep.

Appendix A

Suicide Awareness Information

SIGNS OF SUICIDE

A person contemplating suicide—

- Believes he or she is in a hopeless situation.
- Appears depressed, sad, tearful; may have changes in patterns of sleep and/or appetite.
- May talk about or actually threaten suicide, or may talk about death and dying in a way that strikes the listener as odd.
- May show changes in behavior, appearance or mood.
- May increase or start drug or alcohol use.
- May injure self or engage in risky behavior.
- Abandons planning for the future.
- May start withdrawing from others, including family and close friends.
- May give away possessions.
- May appear apathetic, unmotivated, indifferent.

PREVENTION KEYS

The small-unit leader may be the first to identify the "signals" of a potential suicide-risk Service member. The keys to prevention are to provide aid to persons at suspected risk and follow the acronym AID LIFE.

A	Ask: "Are you thinking about hurting yourself?"
I	Intervene immediately.
D	Do not keep a secret concerning a person at risk.
L	Locate help (NCO, chaplain, corpsman, doctor, nurse, friend, family, crisis line, hospital emergency room).
I	Inform your chain of command of the situation.
F	Find someone to say with the person. Do not leave the person alone.
E	Expedite! Get help immediately. A suicidal person needs the immediate attention of helpers.

Appendix B

Religious Ministry Team's Role

(Adapted from Army Field Manual (FM) 22-51, *A Leader's Manual for Combat Stress Control.*)

This appendix addresses the general role of the ministry team in the commander's program of combat stress control and in battle fatigue ministry. In the Marine Corps ministry teams are called Religious Ministry Teams (RMTs) and consist of chaplains and Religious Program Specialists (RPs) or chaplain's assistants (In the Army they are called Unit Ministry Teams). The RMT is assigned to a command or designated by higher headquarters to be responsible for the direct religious ministry support to the command. RMTs provide professional ministry support to leaders in fulfilling their combat stress identification and intervention responsibilities. RMTs can also assist in training leaders to recognize combat stress symptoms.

EFFECTS OF STRESS

Negative effects of stress can be lessened when members are prepared physically, emotionally, and spiritually prior to combat/operations. RMTs can assist in preparing Service members to manage combat stress with training before and during deployment. This training helps Service members draw upon their personal religious faith, spiritual strength, and values as well as to share strength and confidence during any operation.

RELIGIOUS MINISTRY TEAM RELATIONSHIP

The RMT's established relationship with the command promotes trust with the troops. As integral members of their commands, RMTs are trained and ready to respond to the needs of Service members experiencing combat stress. A person-oriented resource, the chaplain and RP team minister to combat stress casualties, especially Service members having less severe difficulties who have rapid replenishment potential.

SPIRITUAL VALUES

Service members' inner resources are often based on their personal faith, religious beliefs and spiritual values. In combat, Service members may show more interest in their religious beliefs. When religious and spiritual values are challenged during the chaos of combat, Service members may lose sight of inner resources that sustain them. Service members can then become targets of fear, despair, hopelessness, and eventually, combat stress casualties. They are also at risk for committing misconduct stress behaviors. Chaplains are a source of direction and stability to Service members experiencing these dilemmas and seeking to refocus their personal beliefs and spiritual values.

RELIGIOUS MINISTRY TEAM SUPPORT

Ministry teams can provide preventive, immediate, and replenishing spiritual and emotional support and care to Service members experiencing combat stress.

Preventive

The RMT's religious ministry support assists in preventing combat and misconduct stress behaviors through establishment of a presence within the unit. It is important for the RMT to be present with Service members when the unit trains and when it deploys. The chaplain and RP can be a calming influence on members and can help Service members strengthen or regain values important to them. Chaplains can help prevent combat stress and misconduct by:

- Being present with the Service members and deploying with the unit.
- Providing opportunities for private and group prayer and worship.
- Supplying personal religious articles and materials.
- Reading the scriptures with Service members.
- Providing sacraments as the situation allows.
- Counseling Service members, allowing them to work through stress, fear, anxiety, anger, and frustration.
- Visiting Service members in work and living areas.
- Assisting Service members and families prior to deployment with preparation for geographical separation and an uncertain future through Marine Corps Family Team Building (MCFTB) programs that emphasize family strengths. This helps Service members to know that their families are cared for during deployment.

Immediate

RMTs can assist commanders in the identification of Service members experiencing combat stress. Chaplains often work closely with the medical officer and are trained to recognize the signs of combat stress and provide religious support to Service members experiencing combat stress. Chaplains assess spiritual needs and provide the appropriate religious ministry. This may include such things as:

- Presence with the Service member.
- Conversation and counseling, providing opportunities to share fears, hopes, and other thoughts.
- General prayers; prayers for the individual or fallen comrades.
- Rites, sacraments, and ordinances as appropriate.
- Reading from scriptures.

Replenishing

Following an engagement, the unit may require reconstitution through the addition of new personnel. The RMT can assist surviving Service members rebuild their emotional, psychological, and spiritual strength. During this time, RMTs continue their ongoing direct religious ministry that includes:

- Coordinating the availability of worship services, sacraments, rites, and services and/or ceremonies honoring the dead.
- Assisting with the integration of personnel replacements.
- Enabling the grief process through personal counseling and memorial services.

- Reinforcing the Service member's sense of self-worth and hope.
- Assisting with CEDs to provide opportunities for Service members to talk about their combat experiences and to facilitate integration of the combat experience into their lives.
- Providing leadership training and supervision of CEDs.
- Providing personal religious articles and materials.
- Providing or arranging faith group coverage in the command or for other units.
- A chaplain's ability to relate the spiritual aspects of life to the Service member's situation is an essential element in the replenishment process. Religious ministry can assist Service members in achieving emotional and spiritual wholeness.

Reintegration

After deployments and operations, RMTs can assist reintegration of the individual Service member into family relationships and society at large. Many religious ministry support programs remain the same. Expanded ministry may also include providing the following:

- Worship events for the entire unit.
- Worship events for varied faith groups.
- Briefings that help Service members recognize, prepare for, and master the stressors of reunion with family.
- Structured events to assist Service members returning to family and civilian life.
- Opportunities for Service members to experience and understand the forgiving and unchanging love of God.

Appendix C

Glossary

Section I. Acronyms

AAR	after action review
BAS	battalion aid station
BICEPS	brevity, immediacy, centrality, expectancy, proximity, and simplicity
CED	critical event debrief
CSC	combat stress control
MCFTB	Marine Corps Family Team Building
MCRP	Marine Corps Reference Publication
MOPP	mission-oriented protective posture
MRE	meals, ready to eat
NCO	noncommissioned officer
NGO	nongovernmental organization
PVO	private voluntary organization
RMT	religious ministry team
RP	religious program specialist
SGLI	Servicemen's Group Life Insurance
SPRINT	Special Psychiatric Rapid Intervention Team
UMT	unit ministry team
WIA	wounded in action

Section II. Definitions

circadian rhythm—Rhythmic variations in individuals; predictable physiological and behavioral cycle of about 24 hours.

combat stress—The mental, emotional, or physical tension, strain, or distress resulting from exposure to combat-related conditions. Combat stress is the net result of many stressors (e.g., fear of death, fear of failure, other intensive painful emotions [like grief], uncertainty, boredom, worry about what is happening back home, and the many physical and mental demands of combat duties) at any given time.

combat stress behaviors—The generic term that covers the entire range, from highly positive to totally negative behaviors which can be elicited by combat stress.

combat stress reaction—The common, predictable, negative, emotional and physical reactions of normal people to the "abnormally high" stressors of combat, which by definition interfere with mission performance or well-being, but which can be treated according to the BICEPS principles, with reassurance of normality and recovery, with rest, replenishment, and activities to restore confidence. The combat stress reactions do not include the misconduct stress behaviors, since the latter require disciplinary action.

continuous operations—military operations with many pulses of action every day and night, continuing for several days to weeks, which require careful planning and resource allocation to give everyone a minimum of 4 hours sleep in 24.

stress—An internal process of an individual for preparation in dealing with, or reacting to stressors. Stress involves physiological stress reflexes, which ready the body for fight or flight, such as increased release of adrenaline, heart rate, and blood flow to brain and muscles, with subsequent increase in alertness. Stress also involves the accompanying emotions (stress responses) such as anxiety, anger, or guilt. Stress may or may not involve conscious awareness of the threat, but the stressor must be perceived at some level of awareness. The amount of stress experienced depends on the appraisal of the stressor and its context by the individual, even if that perception is incorrect. The stress process includes psychological defenses which may shield the individual from perceiving more threat than he can tolerate.

stress behaviors—Stress behaviors are the observable behaviors that an individual exhibits as a result of internal stress. They may be intended to overcome and turn off the stressor, escape it, adapt to an inescapable stressor, or make the uncomfortable stress symptoms more tolerable. A stress behavior may simply be a side effect of the tension, which is part of stress. Any of these different types of stress reactions may be successful, unsuccessful, adaptive, or maladaptive. They may make the situation better, worse, or not influence it at all. They may resolve one stressor but create new ones.

stressor—Any event or situation that requires a non-routine adaptation or behavior change, creates conflicting motives within the individual, or is perceived by the individual as a threat to his well-being or self-esteem. Combat stressors are any that occur in the context of performing the combat mission (whether under fire or not).

sustained operations—Continuous operations or combat with opportunity for less than 4 hours sleep per 24 hours for significant personnel, which may be brief or fragmented.

Appendix D

References

DoD Directive

6490.5, *Combat Stress Control (CSC) Programs*

Army Field Manual (FM)

8-51, *Combat Stress Control in a Theater of Operations, Tactics, Techniques, and Procedures*

22-9, *Soldier Performance in Continuous Operations*

22-51, *Leaders' Manual for Combat Stress Control*

22-100, Army *Leadership*

90-22, *(Night) Multiservice Night and Adverse Weather Combat Operations*

Miscellaneous

U.S. Army Aeromedical Research Laboratory and U.S. Army Safety Center, *Leader's Guide to Crew Endurance*

http://www.mcu.usmc.mil/TbsNew/Coverpag.htm, *Human Factors in Combat: Topics in Combat Leadership that Help Preserve the Fighting Spirit and Prevent Psychiatric Casualties*

Made in the USA
Monee, IL
05 November 2026

67808315R00157